PRODUCTION OF POT ROSES

H. Brent Pemberton
Texas A & M University
Agricultural Research and Extension Center
PO Box E
Overton, Texas 75684

John W. Kelly
Department of Horticulture
Clemson University
Clemson, South Carolina 29634

Jacques Ferare
The Conard-Pyle Company
372 Rose Hill Road
West Grove, Pennsylvania 19390

Growers Handbook Series
Volume 7

Allan M. Armitage, General Editor

PRODUCTION OF POT ROSES

H. Brent Pemberton, John W. Kelly,
and Jacques Ferare

TIMBER PRESS
Portland, Oregon

Front cover: A well-grown short-cycle crop ready for marketing (photograph by H. Brent Pemberton).

Printed in Hong Kong

ISBN 0-88192-379-6

TIMBER PRESS, INC.
The Haseltine Building
133 S.W. Second Avenue, Suite 450
Portland, Oregon 97204, U.S.A.
1-800-327-5680 (U.S.A. and Canada only)

Library of Congress Cataloging-in-Publication Data

Pemberton, H. B.
Production of pot roses / H. Brent Pemberton, John W. Kelly, and Jacques Ferare.
p. cm. — (Growers handbook series ; v. 7)
Includes bibliographical references (p.).
ISBN 0-88192-379-6
1. Roses. 2. Rose culture. 3. Forcing (Plants) 4. Plants, Potted. I. Kelly, John W. II. Ferare, Jacques. III. Title. IV. Series.
SB411.P38 1997
635.9'33372—dc20 96-27599
CIP

Contents

1 Introduction

The growing of roses in pots is not new! As early as 1869, elaborate instructions were available concerning the forcing of roses out of season in large containers (Parsons, 1869). By 1949, techniques had sufficiently advanced that Dr. Kenneth Post of Cornell University described the use of roses for pot culture for the Easter and Mother's Day holidays (Post, 1949). Traditional spring pot forcing of field-grown bare-root grafted plants of the floribunda and polyantha type reached its peak after the second world war, but by the 1970's this market was in decline. Efforts were made to expand bare-root pot rose forcing to include the Valentine's Day holiday through the use of early field digging and supplemental light (Heins, 1981), but the practice was never adopted on a large scale in the United States due to economics.

Based on the work of Moe (1973), growers in Europe developed an economical system of year-round production in pots employing the use of cutting propagation and supplemental light. This system also used a smaller pot than traditionally used for bare-root grafted plant forcing, which was more suited for mass market sales. Beginning in the 1980's, new miniature rose cultivars were introduced that exhibited greenhouse and postharvest performance superior to those traditionally used. In recent years, the accelerated introduction of new cultivars and improvements in greenhouse technology have resulted in increasingly efficient year-round production schedules. The new cutting-grown products have been widely accepted by consumers, resulting in a significant expansion of the pot rose market. Also, recently introduced cultivars have been used for spring forcing of bare-root grafted plants, resulting in a rejuvenation of the traditional market.

The impact of the new cutting-grown products has been dramatic on the European market over the past decade and, in the last few years, has been felt in North American markets as well. Current production in Europe is estimated at more than 50 million pots. The majority of this production is in Denmark (35 million) and Holland (10 million), with the remainder grown in France, Germany, and Italy. U.S. and Canadian production was estimated at 4 million pots in 1989 but is estimated at over 12 million for 1997. Most production is in 4 to 5-inch (10 to 12.5-cm) pots, but there is a tendency for using larger, 5 to 6-inch (12.5 to 15-cm) pots in North America and smaller, 2.5 to 3-inch (6 to 7.5-cm) pots in Europe.

The objective of this production guide is to present our current state of knowledge of pot rose production for use by growers, researchers, and educators. Both traditional and recent advances in production methods are included. Pertinent examples from the scientific literature have been included to form a basis of understanding of how rose plant physiology can be used to improve production techniques. The intent is to stimulate interest in pot rose production research and to improve the efficiency of production and postharvest handling so that a growing number of consumers can enjoy this beautiful and unique plant.

2 Selection of Cultivars

AVAILABLE CULTIVARS

Traditionally, cultivars from the polyantha and floribunda classes have been used for pot rose forcing (Table 1). Dick Koster and Margo Koster (polyanthas) have been used extensively since their introduction in 1929 and 1931, respectively. The floribunda Garnette and its many sports (e.g. Carol Amling, Marimba, and Bright Pink Garnette) were introduced in the 1950's and 1960's. Until recently, they were still among the mainstays of the pot rose forcing industry in North America, especially for larger containers (Hammer, 1992).

Use of cultivars from the miniature class is increasing rapidly (Table 1). Traditionally, numerous cultivars of miniatures were available, but few were suitable for pot forcing due to poor shelf-life or the inability to provide uniform growth and flowering on a year-round basis. Due to recent breeding breakthroughs, however, the palette of cultivars for pot forcing has rapidly broadened (Plate 1). Many new cultivars have been introduced from Europe where they have become an established floricultural crop, but North American breeding programs also continue to expand. Most modern pot rose production is and will continue to be with miniature-type cultivars.

Two main groups of miniature cultivars (all classed as miniatures) suitable for pot forcing are the true miniatures and the large-leaf/large-bloom types (Table 1). The true miniatures are most suitable for a 4-inch (10-cm) pot program because they are compact, well branched, heavy flowering, and may be forced under low light intensity. Compared to the larger miniatures, the true miniatures are usually easier to grow and produce more, but smaller, flowers. True miniatures are well adapted for a Valentine's Day crop, but are also suitable as an impulse product year-round. They bloom continuously throughout the growing season in the garden; however, they may not be winter hardy in the northern United States and Canada. The best known cultivars are from the Rosamini, Minimo, Parade, Rainbow, and Mini-wonder series.

The large-leaf/large-bloom types are the likely successors of the Garnettes and Kosters. They are smaller, better branched, and more compact and uniform in habit than the Kosters. The vividly colored flowers, which are smaller than the floribundas but larger than the true miniatures, are highly resistant to shattering. Plants are best adapted to a 4.5 to 6-inch (11 to 15-cm) pot program. They are also available as budded, field-grown plants (see Chapter 4), which are forced in 7 to 8-inch (18 to 20-cm) pots similar to the Garnettes and Kosters. Most of these cultivars exhibit good winter hardiness and disease tolerance. Cultivars of the Sunblaze series are good examples of the large-leaf/large-bloom miniature rose.

The above-mentioned miniature cultivars have been selected by rose breeders specifically for greenhouse production. Other miniature cultivars might also be suitable for greenhouse forcing, but selection must be made from the hundreds of cultivars presently available for garden use. While dormant plants of garden cultivars can be successfully forced in the spring and sold for garden use, cultivars suitable for pot-plant use must exhibit uniform growth and flowering after a severe pinch (pruning) on a year-round basis. A good shelf-life is essential for marketing and indoor enjoyment.

THE SELECTION PROCESS

New pot rose cultivars are being developed through the almost exclusive use of genetic material from the miniature class in modern breeding programs. Because they are relatively easy to breed, large quantities of miniature rose cultivars are available today; however, only very few qualify as "pot roses." An intensive and thorough selection process is used to determine the suitability of a rose selection for pot-plant production. Some of the important attributes are the following:

- At the production level, cuttings must root readily and uniformly; plants must exhibit uniform, well-balanced growth with heavy flowering and be easy to schedule, quick to cycle from pinching or cut-back (heavy pruning) to flower, and disease tolerant.
- At the distribution level, plants must withstand adverse long-distance shipping practices; flowers must be persistent and have a strong, stable, attractive color.
- At the consumer level, plants must have a persistent flowering period and be easy to maintain for their intended use, whether indoors, outdoors, or both.

In order to meet these criteria, pot rose cultivars usually require 5 to 7 years of evaluation before they are released. Most of the cultivars available today have been rigorously selected for pot-plant culture, especially the more recently introduced ones. Breeding of pot roses for the characteristics described above is taking place in Europe, Canada, and the United States, bringing new colors and more adaptable plants to the marketplace each year.

A typical breeding and selection program begins with the crossing of parents determined by the breeder to have the characteristics desired and a high probability of being able to pass the genes for those characteristics to the resulting progeny. The selection process then follows and usually consists of 5 main stages of development from the time the seedlings germinate until a new cultivar is released.

Stage 1

The first selections are made for color and habit from hybrid seedlings, which may number in the thousands.

Stage 2

From the first selection, typically about 10% of the seedlings may show desirable characteristics and are propagated by cuttings for further testing. This stage lasts 18 to 24 months, involves several cycles of growth, and focuses primarily on plant performance characteristics such as habit, floriferousness, ease of propagation, and forcing capacity.

Stage 3

At this stage, the best seedlings are observed in the greenhouse under standard commercial production scheduling. Cutting-grown plants are used for greenhouse testing that focuses on production timing, shipping tolerance, and shelf-life. Some companies graft all selections on a rootstock at this stage to evaluate field and garden performance. Usually, less than 50 selections reach this stage in any given year.

Stage 4

After year-round testing on greenhouse performance, shipping and shelf-life tolerance, and garden performance (Stage 3), no more than 6 to 10 selections are likely to remain. By this time, the technical characteristics of the cultivars are well defined. The potential market value is then evaluated by selected growers who grow the new selection under "real production conditions."

Stage 5

Following a successful market trial, a selection will be named and the new cultivar may be patented and released to the trade. Because of the extensive amount of testing required to develop any new pot rose cultivar, virtually all the selections deemed acceptable in Stage 4 are protected by plant patent. Therefore, a grower interested in propagating and growing plants of these cultivars must first obtain a propagating license from the patenting company.

VALUE AND IMPORTANCE OF PLANT PATENTS

New, asexually propagated cultivars are usually protected by plant patent or plant breeder's rights in most countries where pot roses are grown. Licensed growers of patented cultivars pay a royalty, usually per plant sold or per plant rooted, to the breeder and/or patent holder. Once granted, a patent protects a cultivar for 17 years in the United States. At this time, plant patents are recognized in the country of origin only, but a cultivar can be patented in other countries as well. The introduction of patent laws provided plant breeders the right to be rewarded for their inventiveness. Growers who pay royalties on patented cultivars are, in fact, providing funds to support research programs that are both costly and long-term. These programs, in turn, benefit growers by supplying them with cultivars exhibiting new colors, better disease resistance, better greenhouse performance, and improved shipping capacity and shelf-life. Without plant patents, little or no incentive exists to invest the large funds and resources needed to create new cultivars.

A WORD ABOUT NOMENCLATURE

Currently there is much confusion over botanical nomenclature in roses. Traditionally, a name was assigned to a rose and registered with the American Rose Society (ARS), the international registration authority for roses (Cairns, 1993). In this way, the assigned name became the cultivar name for that rose according to the international code of nomenclature (Brickell, 1980). However, modern practice has evolved into assigning a code (denomination) name to a breeding selection prior to introduction (e.g. Meijikatar). This name is used for patenting, which is typically applied for before the rose is introduced commercially. A common name (e.g. Orange Sunblaze) is chosen and usually trademarked for marketing purposes when a rose is introduced commercially. Both the denomination and common names can be registered with the ARS. Since both names can be registered, resolving which is the correct cultivar name is problematic. However, the denomination name will always be the same for a specific rose, as will the patent number, because it was used in the patent application. The denomination name is generally considered to be the correct cultivar name, especially when the common name is trademarked. According to the code of nomenclature, the cultivar names should not be trademarked (Brickell, 1980).

Another controversy exists concerning the use of trademarked names for selling roses. Many view the use of trademarked names as a means of circumventing the expiration of a patent because trademarks are renewable whereas plant patents are not (Higginbotham, 1992). After a patent expires, anyone can legally propagate a rose, but it cannot be sold without permission under the name that was trademarked by the company originally holding the patent. Nevertheless, selling roses under trademarked names is common practice. There is disagreement, however, over whether trademarks are being used correctly. Elliot (1991) states that using a trademark to market a single variety is not prevented by law. However, improper use can easily invalidate the trademark. For this reason, trademarked and cultivar names are commonly used together. If they are not, there is a danger that the trademark can be genericized, thus rendering it invalid (e.g. if a rose is only known in commerce by the trademarked name) (Elliot, 1991; Higginbotham, 1992).

There are commonly many conflicting opinions on the naming of roses and use of trademarks (Higginbotham, 1992). Revisions to the code are currently being considered to solve these problems, but cooperation and compromise will be needed to end the confusion (Higginbotham, 1992). It is hoped that these issues can be resolved to the benefit of scientists, plant breeders, plant producers, and consumers alike.

For the purposes of this book, the presentation of rose names has been simplified. Single quotes, typically used to denote cultivar names, and marks denoting trademarks of any kind have been omitted in the text. Names in common use in North America have been used when referring to research originally reported using correct synonyms. Table 1 contains information on all of the pot rose cultivars referred to in the text, and many of those available in commerce. This table also lists information concerning common names, code names, synonyms, patents, and claims of property rights concerning names and propagation.

3 Forcing Strategy

The specific plant-production and forcing methods used for pot rose production are dependent upon the choice of forcing strategy. Details of the propagation and production of plants for forcing are in Chapter 4, and details for the forcing phase of pot rose production may be found in Chapter 5. The choice of forcing strategy is driven by marketing requirements which not only dictate cultivar selection, but time of year for forcing and pot size as well (see Chapter 9). Plants produced for forcing are either field-grown bare-root grafted plants, bench-grown bare-root rooted cuttings, or container-grown as liners or pre-finished units.

Field production is used to produce dormant, grafted, bare-root plants for forcing (Plate 2). Because of the typically large size of these plants, they are usually potted into 6.5-inch (16.5-cm) or larger pots for forcing. Field-grown plants are only available for a limited time during the winter and spring and are typically used for Easter and Mother's Day forcing. Forcing time is 6 to 8 weeks. Forcing of field-grown plants has been traditionally used to produce crops of the larger flowered cultivars such as the Garnettes and Kosters, but is also now used to produce plants of the large-flowered type miniature cultivars such as the Sunblaze series and specialty items such as patio trees (Plate 3).

Another type of bare-root plant that is produced for forcing is the dormant bare-root rooted cutting. These are potted into 4 to 5-inch (10 to 13-cm) pots with availability and forcing time the same as for field-grown plants. These plants are considerably smaller than field-grown plants, however, because this procedure is primarily used to produce plants of the small-flowered miniature cultivars for small pot forcing.

SHORT-CYCLE WITH DIRECT STICKING

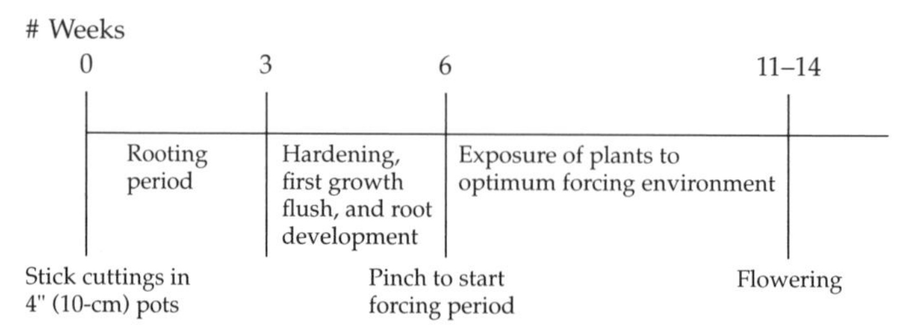

SHORT-CYCLE WITH LINERS

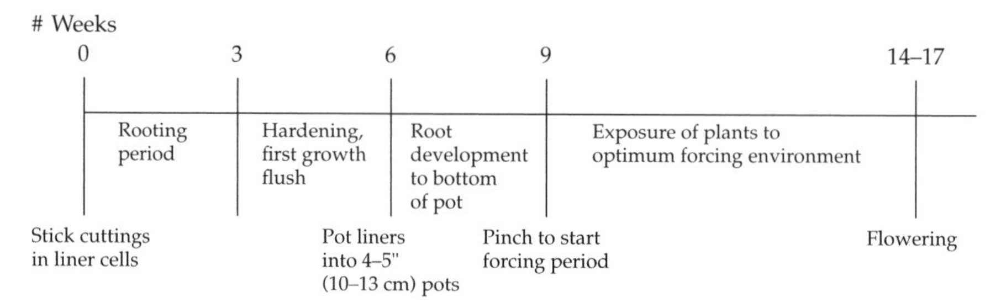

Figure 1. Examples of short-cycle crop-production schedules. See Chapters 4 and 5 for details of propagation and forcing techniques.

Potted roses grown in containers from start to finish are produced by using the smaller flowered cultivars with 1 of 2 methods, the short-cycle or the long-cycle method. For the short-cycle method, actively growing liners or pre-finished plants are commonly used (Figure 1 and cover photo). Depending on plant size, liners can be potted into pots 4 to 6 inches (10 to 15 cm) in diameter. Plants pre-finished in 4-inch (10-cm) pots can be either forced as is or finished in 6-inch (15-cm) pots. Many larger producers propagate directly into the pot to be used for finishing (referred to as direct-sticking; Figure 1). Regardless, the key to success with the short-cycle is allowing the roots to become well established in the finishing pot. This usually takes 2 to 3 weeks for newly potted liners. The plants are then pinched to 2 inches (5 cm) above the pot rim to start the forcing period. Forcing time is 5 to 8 weeks depending on time of year and local conditions (see Chapter 5). If plant roots are not well established when pinched, poor and uneven bud-break results in a finished plant with poor balance. This is the most commonly used forcing method for finishing plants from March to November in North America. In northern Europe, supplemental high intensity discharge (HID) lighting and CO_2 are added to produce plants year-round. Winter forcing and year-round production with supplemental lighting are also common in the U.S. and Canada.

With the long-cycle method, better results are obtained when forcing for winter and Valentine's Day flowering (Plate 4). Plants used for long-cycle forcing are container grown during the summer and allowed to go dormant in the autumn (Figure 2). Plants are usually ready to force by early December in most areas of North America. To force, the plants are cut back 2 to 3 inches (5 to 7.5 cm) above the pot rim, cleaned of all dead branches and debris, and put in the forcing environment. Uniform bud-break and growth is achieved by providing a good balance between light and temperature (see Chapter 5). Seven to 8 weeks are usually needed to finish long-cycle plants for Valentine's Day or 6 weeks if HID supplemental lighting is used. The long-cycle method results in plants that are woody and well branched with a high

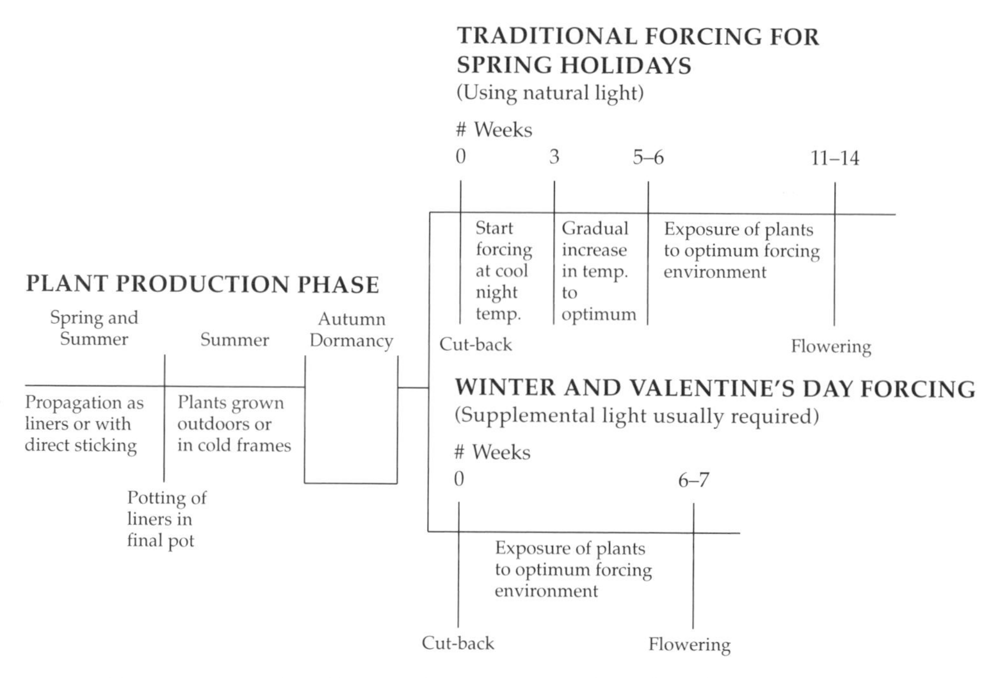

Figure 2. Example of long-cycle crop-production schedules. See Chapters 4 and 5 for details of propagation and forcing techniques.

carbohydrate content. The well-developed buds and root system (Plate 5) result in plants with vigorous growth that finish under the lower light intensity of the winter months significantly better than a short-cycle plant. In addition, the long-cycle plants can be used to force for Easter or Mother's Day in less expensive growing structures such as cold frames. However, the availability of long-cycle grown plants of good forcing cultivars is currently limited in the U.S. Long-cycle production is more common in large nurseries where the plants are used internally rather than sold to finishers. These operations often use long-cycle grown plants to increase the number of plants available for finishing for major winter and spring holidays. When year-round, short-cycle production is sustained with plants grown with cuttings from each final pinch, the number of plants that can be produced on a weekly basis is limited by the number of cuttings that can be generated from existing crops. Because this number is not generally much more than that necessary to sustain current production numbers, increases can only be accomplished gradually. By utilizing excess cutting numbers from short-cycle crops during late spring and summer to produce long-cycle plants, large numbers of plants can be finished for subsequent peak holidays such as Valentine's Day and Mother's Day.

4 Production of Plants Used for Forcing

FIELD PRODUCTION OF GRAFTED BARE-ROOT PLANTS

Field production of dormant, bare-root plants for later greenhouse forcing in pots is centered in California. A 2-year production cycle is generally used for producing grafted plants of polyantha and floribunda cultivars. The first step in plant production is the removal of hardwood rootstock cuttings from established crop rotations. All but the apical 2 or 3 buds are removed from each cutting prior to planting in late autumn or early winter. The following spring, a bud from a desired cultivar is T-bud grafted onto the rootstock stem just below the actively growing shoots. The grafted scion bud is usually inactive until late winter of the following year, when the rootstock top is removed. Active growth of the scion bud commences and continues through the second growing season. The dormant plants are harvested during late autumn and early winter at the end of the second year of production. Detailed cultural practices are presented elsewhere (Pryor et al., 1987).

A 1-year production cycle similar to that used to produce "started eyes" for the greenhouse cut flower industry (Pryor et al., 1987) is used for growing grafted plants of the large-bloom type miniatures (Plate 2). The cultural procedures are similar except that the rootstock top is broken, but left partially attached, just above the graft union after the T-bud graft has been allowed to heal (about 21 days). When the scion begins flowering for the first time, the rootstock top is completely removed. Dormant plants are then harvested at the same time as the 2-year-old bush roses.

Miniature standards, referred to as mini- or patio trees, are also produced by using a 1-year production cycle. Disbudded, 24-inch

(60-cm) rootstock cuttings are planted 6 inches (15 cm) deep at the same time as for bush roses, covered with white plastic sleeves to control moisture loss during rooting, and tied to supports. The following spring, each rootstock is T-bud grafted 4 times with the desired scion at a height of 19 inches (47.5 cm) in order to ensure a well balanced, uniformly flowering canopy. After the grafts have healed (approximately 21 days), the rootstock top is removed just above the graft union to force the scions into growth. Dormant plants are then harvested at the same time as the bush roses.

Choice of rootstock for grafted plant production is primarily based upon tradition, grower experience, and availability rather than any scientifically demonstrated advantage. *Rosa multiflora,* Dr. Huey, and Manettii are used for 2-year plant production; Manetti is used for 1-year production of large bloom type miniatures and Dr. Huey is used for mini-tree production. Scions of Mothersday budded onto *Rosa multiflora* have been found to grow more vigorously than scions budded onto *R. multiflora* cv. Japonica (Moe, 1970), indicating that rootstocks could be selected for growth control (Hammer, 1992).

When plants are harvested, they are mechanically defoliated, pruned according to grading standards (see below), and the canes dipped in a fungicide (Hardenburg et al., 1986). They are placed into cold storage at 31 to 36°F (–0.6 to 2.2°C) and sprayed frequently with water to maintain 85 to 95% relative humidity until needed for potting. Plants are packed in large polyethylene-lined boxes for refrigerated shipping to forcers. They are graded according to 2 systems for marketing. One system consists of 3 grades: "X"—at least 1 strong cane; "XX"—at least 3 strong canes; or "XXX"—at least 4 strong canes (Hammer, 1992). The other system consists of numbered grade standards established by the American Association of Nurserymen (Table 2; Anonymous, 1990).

PLANT PRODUCTION FROM CUTTINGS

Cutting propagation is generally used to produce plants of all cultivar types from the miniature class (see Chapter 2). The key to efficient production of asexually propagated cuttings is the use of "clean" stock-plant material, free of pests and diseases. Cuttings can be obtained either from blocks of stock plants maintained specifically for cutting production or from the shoots obtained from a short-cycle crop when it is pinched to start the final forcing period. Commercial propagators use the first option, whereas large growers involved in a continuous, year-round growing cycle tend to favor the second, especially in modern, fully controlled greenhouse environments. Many smaller and/or seasonal producers purchase rooted cuttings from a specialist propagator. In this case, obtaining the highest quality product possible on a reliable basis becomes the main concern. Another consideration when deciding between cutting propagation and purchase of rooted cuttings is that a license is required to propagate patented cultivars. Royalties are collected based upon the number of plants successfully propagated and sold. Propagation of patented cultivars without a license is illegal.

Cutting Production from Stock Plants

Many factors affect cutting production from stock plants, but any condition promoting healthy growth of rose plants should enhance cutting production. Specific recommendations for cutting production are presented here. Additional details of how these various factors affect rose growth can be found in Chapter 5.

Stock-plant Selection and Establishment

To assure the maximum production of cuttings, great care must be taken in choosing the best plant material possible to use as stock plants. Too often, growers select stock plants from "left-over" plants that could not be sold out of a forced crop. Most likely, such plants are either diseased or below grade. By starting with low-quality plants and by repeating this process time after time, the grower, in fact, promotes a negative selection process. The quality and productivity of the crop become worse over time. To avoid this very costly process, only the best cuttings from the best plants must be selected to start new stock.

Stock plants should be well established before the first cuttings are harvested to obtain maximum cutting production. Any stress during the first 2 to 3 weeks after repotting should be avoided. Water, temperature, or mechanical stress (such as pinching too early) during that stage greatly reduces the plant's potential to produce cuttings. Although the plant usually recovers, these stresses delay its normal development and in some cases can be fatal.

The yield of cuttings is directly related to the number of buds breaking on the stock plants. Maintenance of young, vigorous growth is critical because buds on older, woodier canes take longer to break, and fewer breaks occur than on softer, younger canes (Khayat and Zieslin, 1982). To accomplish this in a stock block, plants are maintained no more than 1 year in the same container. Containers larger than 6-inch (15-cm) pots should be used to maintain root activity at optimal levels during the entire cutting production cycle. Repotting to a larger container once a year is also possible. Plants should be spaced approximately 1 per square foot (10.8 per square meter) so that the canopies do not overlap. When stock plants in containers are too old, their cutting production drops drastically. They are then pruned very heavily to 4 to 5 inches (10 to 13 cm) above the rim, and can be sold as finished plants on the next flush of flowering.

Light

Light intensity is the most important factor affecting rose plant growth (Zieslin and Mor, 1990). The importance of good light conditions for cutting production cannot be overemphasized. Because of this, most propagators and year-round producers are located in high-light areas such as Florida, Texas, southern California, Israel, and southern Europe, or use artificial lighting to complement natural radiation as in most of the U.S., Canada, Holland, and Denmark. Light intensity below 800 footcandles (105 µmol per second per square meter of high pressure sodium light; see Table 6) results in poor breaking and production of small shoots that do not root or grow with the vigor needed for commercial production.

Long photoperiods have also been found to stimulate growth and flowering of rose plants in addition to that obtained by maintaining adequate light intensity (Zieslin and Mor, 1990). However, incandescent lighting has been shown to reduce bud break (Moe, 1972), whereas high pressure sodium or fluorescent lighting has been shown to promote bud break (Carpenter and Anderson, 1972; Khosh-Khui and George, 1977). Day extension to at least 16 hours with high-pressure sodium lighting at a minimum intensity of 500 footcandles (66 µmol per second per square meter) is recommended for maximum cutting production.

Temperature

Night temperatures below 60°F (15.5°C) result in poor bud break and day temperatures above 85°F (29.5°C) reduce the quality of cuttings. A range of 62–65°F (16.5–18.5°C) night and 72–75°F (22–24°C) day temperature is most commonly used in the United States and Canada. However, growers in northern Europe are successfully using a uniform day and night temperature of 70–74°F (21–23.5°C) to produce high quality pot rose cuttings.

Carbon Dioxide

Carbon dioxide enrichment is highly beneficial to rose growth in general (Mastalerz, 1987). A level of 1000 to 1500 ppm increases bud break and fresh weight of new shoots on cut-flower cultivars compared to a non-enriched atmosphere. CO_2 enrichment did not increase the number of flowers or flowering shoots on Orange Sunblaze plants, but did increase height and dry weight of the shoots (Clark et al., 1993; Mortensen, 1991). This effect was enhanced by increased light levels.

Growing Media and Fertilization

A well-aerated growing mix with good drainage facilitates the availability of oxygen to the root system and is necessary to achieve vigorous plant growth and thus efficient cutting production. A heavy, poorly drained growing medium inhibits root development and activity and decreases rapid, uniform bud breaking and cutting production. Incorporating superphosphate and dolomitic limestone in the mix helps balance the pH and provides calcium and magnesium. The optimum pH for pot rose production is 5.5 to 6.0. A pH above 7.0 results in chlorosis and poor quality cuttings.

Because pot roses are grown under a wide range of climatic, environmental, and geographical conditions, recommending a specific fertilization program suitable to all growers is impossible. However, most growers use a moderate constant-feed program of 150 to 180 ppm nitrogen. This is applied with a 2-feed, 1-leach schedule whereby every third watering is done without fertilizer to prevent soluble salt accumulation. Any good, commercially available fertilizer such as 20–20–20 or 15–10–12 or equivalent can be used. Fertilizing stock plants with excessive nitrogen levels (above 300 ppm) on a continuous-feed program can negatively affect rooting of the cuttings taken from them. Iron chelate applied at 5 ppm weekly will insure vigorous growth and improve leaf color if needed. Micronutrients should be incorporated in the feeding program to insure proper growth.

Use of soil and foliar sample analysis on a regular basis is a helpful tool to monitor stock-plant fertilization. Media/soil samples should be taken regularly in order to check pH and nutrient levels. Foliar analysis should be done regularly to check micronutrient levels (see Chapter 5).

Water

Any condition leading to loss of roots and/or foliage will decrease the vigor and cutting production of a stock plant. Either too much or too little moisture can result in leaf yellowing and abscission and root loss (White and Holcomb, 1987). Care must be taken not to overwater newly repotted or planted stock plants during establishment. Also, excessive drying of the media can result in soluble salt injury if fertility levels are high (White and Holcomb, 1987). See Chapter 5 for further details.

Seasonal Variations

The seasonal differences existing in most climates due to light and temperature changes can cause large variations in yield and quality of cuttings at different times of the year. For example, cuttings from Florida in January are usually better than those produced in the northeastern U.S., but the reverse is true in June or July as temperatures become too hot in Florida and favorable temperatures and light conditions prevail in the northeast. To minimize these fluctuations, the grower must try to stabilize environmental conditions throughout the year. For example, 20% to 35% shading combined with fan and pad cooling can be used in hot, high-light areas during the summer to aid in temperature control, whereas supplemental light can be used in low-light areas for winter production. However, seasonal changes in light and temperature control the productivity of rose plants (see Chapter 5). These changes can be dampened by environmental modification but not eliminated (Armitage and Tsujita, 1979).

Pests and Diseases

In managing a block of rose stock plants, *Botrytis* is certainly the most problematic disease and is probably more difficult to control on stock plants than on a finishing crop because of the dense canopy on older plants. It is a particularly critical problem during cool damp weather. The *Botrytis* disease organisms are always present in the environment; therefore, keeping greenhouse humidity low, avoiding free water on the leaves, removing old fallen leaves and debris, and removing the blooms as they open will help control the disease on stock plants.

Powdery mildew is another fungal disease that must be monitored. Adequate spacing to encourage air circulation and control of humidity levels at night help prevent this disease.

Cylindrocladium is a fungus that can result in significant stock-plant losses. Disease occurrence is enhanced by warm, humid, low-light conditions and plant stress. Infected plants should be destroyed at once.

The most damaging pests to stock plants are spider mites. Careful monitoring and spray programs are necessary for control. Any stock plant severely infected by either pests or disease should be immediately removed and destroyed.

More information on pests and diseases affecting pot roses is discussed in Chapter 6.

Cutting Propagation

A well-managed stock-block program should yield at least 20 cuttings per square foot (215 per square meter) per month for large-flowered/large-leaf miniature cultivars and up to 50 cuttings per square foot (538 per square meter) per month with the true miniatures. These numbers represent the average winter production (October to April) over a 5-year period from stock plants grown in 1-gallon (8-inch; 20-cm) pots spaced 1 per square foot (10.8 per square meter) at 62°F (16.5°C) night temperature in southeastern Pennsylvania. Supplementary HID sodium-vapor lighting (350 fc–46 µmol per second per square meter) was also used for 13 hours per day (J. Ferare, unpublished). These data are provided only as a reference, as cutting production will vary significantly with geographical location. Time between cutting harvests is approximately 4 weeks in the summer and 8 weeks in the winter for the northeastern U.S. Growers on a year-round schedule use cuttings from the final pinch of their finishing crop for propagation.

Cutting Type

The ideal cutting for propagation should be taken on newly developed shoots where flower buds are visible but not fully developed (Figure 3). This stage is known as "pea-bud" because the buds are the size and color of small peas. However, cuttings taken at an earlier or later stage will also provide good results if the proper propagation environment is provided.

That pot roses have been selected for easy rooting, node position, number of nodes, and presence of side shoots does not appear to affect the propagation of most cultivars. Single-node, multiple-node, and tip cuttings can be used (Figure 4). Single-node cuttings are mostly used when plant availability is limited or when maximum uniformity is required. Multiple-node cuttings, usually 2 or 3 nodes, are the most widely used either for a direct-stick, 4-inch (10-cm) pot, year-round program (Plate 6) or for rooted-cutting production. Multiple-node cuttings are relatively easy to handle and withstand stress during propagation better than both single-node and tip cuttings. Tip cuttings are taken prior to the visible bud stage, which is an earlier, more succulent stage than single- or multiple-node cuttings. Tip cuttings branch more rapidly than node cuttings but are much more sensitive to moisture stress. They are also more variable in rooting success and subsequent plant development than node cuttings, so the 2 types should not be mixed in the same liner or pot. However, tip cuttings can be used to optimize cutting production, as they can be harvested up to 2 weeks earlier than node cuttings.

Handling the Cuttings

Stock plants should be well-watered and turgid prior to taking cuttings. When harvesting, preparing, and handling the cuttings, clean material, tools, and working areas are very important. Cut-

Figure 3. A flower that is in the "pea-bud" stage.

Figure 4. Three types of cuttings: single node *(top left),* multiple node *(top right),* and tip *(bottom).*

tings should be taken with clean clippers or pruning shears, which should be sprayed often with alcohol or a disinfectant during harvest of the cuttings. Flowers or flower buds should always be removed prior to sticking the cuttings, but the bottom leaf on a multiple node cutting need only be removed if the cut was made just below a bud. Removal of too much leaf area, particularly on single-node cuttings, will inhibit the rooting process (Moe, 1973). Some propagators dip the cuttings in a fungicide solution prior to propagation.

The most damaging problem with pot rose cuttings is improper handling from the time they are removed from the stock plant to when they are stuck in the propagation area. Rose cuttings are very sensitive to moisture stress and high temperature. Experience has shown that drying out or exposure to temperatures above 85°F (29.5°C) for even a short time (30 minutes) drastically increases propagation losses. Stress occurring during the handling of cuttings results in rapid leaf yellowing followed by leaf drop and ultimately the death of the cutting. Leaf loss inhibits the rooting process and stimulates premature bud break, thus resulting in weak growth or loss of the cutting (Moe, 1973). To avoid such problems, the time between cutting and sticking must be kept to a minimum by harvesting only the number of cuttings that can be handled during the same day. Cuttings should always be kept moist and away from high temperatures. They may be stored in a cooler at 35°F (1.5°C) if kept moist, but storage for more than 12 hours is not recommended.

Rooting Hormones

With traditional miniature rose cultivars, the use of a rooting hormone is usually required for optimum rooting (Mor and Zieslin, 1987). However, most pot rose cultivars root without hormones. The majority of the pot roses produced in the greenhouse "plant factories" of northern Europe are propagated without rooting compound. However, in a less controlled environment, the use of a 5-second quick-dip of the freshly severed cutting base in a solution of indolebutyric acid (IBA) at 0.1 to 0.2% is very beneficial (Moe, 1973). Powder applications can also be used. The main effect of IBA is not to increase the total percentage of cuttings that root, but rather to accelerate the process, thus reducing the length of the high-risk phase during rooting. IBA can also increase the number of roots and slow bud break while roots are beginning to grow so that subsequent shoot growth is more vigorous (Moe, 1973).

Media

Pot roses should be propagated in a clean, well-aerated medium free of pathogens. Most commercial propagators use a mix designed to fit their specific needs. Most commonly used are peat-lite-based mixes, but inorganic media such as rockwool or foams such as Oasis have also been used successfully. The propagation medium should always be watered thoroughly before sticking. The most important physical characteristics of a good propagating medium for pot roses is good drainage and aeration. This insures good root development, especially when mist propagation is used.

Drainage can be a problem for growers who direct-stick in the final container, as a mix that is best for propagation is usually too light and does not hold water well enough for the finishing plant. To overcome this problem, growers use a mix that provides a compromise between the requirements of propagation and the needs of the finishing plant. Propagation systems that maintain high air humidity without saturating the medium are advantageous. This balance can be accomplished by the use of fog, tent propagation, or modified high-pressure mist in which the excess water is not able to saturate the pots (Hartmann et al., 1990).

Propagation Environment

As with most other asexually propagated commercial floriculture crops, there are 2 stages in the propagation of pot roses: (1) root initiation and development, and (2) acclimatization and growth. Root initiation and development are critical stages because the cuttings depend on the environment for survival and are extremely vulnerable. As soon as possible after preparation, cuttings should be stuck in a moistened medium and watered in. During this phase, leaves must be kept moist because any moisture stress results in the death of the cutting. This stage lasts 10 to 14 days under normal propagation conditions. At that time, cuttings should be well callused and some new roots should be visible. However, up to 3 weeks may be needed, depending on the cultivar and time of year, for all cuttings to develop visible roots.

During root development, humidity should be reduced progressively to acclimatize the plant to growing conditions. Reducing humidity can be achieved by reducing the duration of misting progressively, gradually opening the plastic when tent propagation is used, or reducing the relative humidity of the fog area. Some propagators also drench with a fungicide at this point to control soilborne pathogens.

How the light intensity required for propagation is maintained depends upon location. Excess or insufficient light can lead to severe leaf yellowing and leaf drop. In northern latitudes, full sun is commonly used, although humidity levels must be sufficiently high to prevent desiccation. At southern latitudes, shade is commonly used to moderate temperature and to cut water loss. Shading of 30% is common, but more than 35% should never be used. A minimum light level of 1500 fc (300 μmol per second per square meter) should be maintained for summer propagation. In areas with low winter light, supplemental lighting may be necessary for winter propagation (Table 3).

The optimum temperature of media for root initiation and development is between 70 and 72°F (21 and 22°C). Temperatures below 68°F (20°C) significantly increase rooting time and decrease rooting percentages.

The earlier in the process that cuttings are fertilized, the greater the fresh weight and overall quality of the rooted cutting. Therefore, a regular fertilization program may be started as soon as the roots begin to develop. Incorporation of a slow-release fertilizer such as Osmocote at the recommended rate is also beneficial. Foliar feeding has been used successfully during the early stage of propagation. One or 2 sprays of 10–52–10 after 1 week has been observed to help promote root development of slow-growing cultivars.

Pot rose cuttings propagated in a CO_2 enriched atmosphere have been observed by northern European propagators to produce more fresh weight, larger leaves, and faster growth than cuttings propagated without CO_2. The concentrations vary from 500 to 1200 ppm, but 800 to 1000 ppm are those most commonly used.

The most important disease during propagation is the soilborne pathogen *Cylindrocladium*. Often confused with *Botrytis*, disease incidence is seasonal (see Chapter 6). *Cylindrocladium* is most active in Pennsylvania in late summer and early fall. In California, it affects crops propagated from May to September more than those propagated during winter. However, the organism is always present and it is difficult to control. The best control is the implementation of sound sanitary practices in propagation, though some fungicides are being tested. Activity appears to be higher under warm, humid, low-light conditions but very little is actually known about how environmental conditions affect *Cylindrocladium* activity on roses.

Outdoor Propagation

At the other end of the spectrum, some growers make use of unsophisticated yet economical propagation techniques and facilities by propagating outdoors during summertime. This method can be successful by slightly adapting the propagation methods described above to the outside environment. Potential problems to consider with outdoor propagation systems include excessive rainfall, impact of wind on the uniformity of mist application, protection from sudden temperature changes and extremes, and increased incidence of disease. A typical program for outdoor propagation and suggested solutions to these potential problems are provided in Table 4.

Finishing of Cutting-derived Plants to be Used for Forcing

Short-cycle, long-cycle, and bare-root rooted cuttings are the types of plants produced from cuttings for forcing. After the propagation phase, rooted cuttings are exposed to various types of production environments, depending on the type of plant to be finished. The effects of various environmental parameters on this phase of production are similar to those described for the forcing phase (detailed in Chapter 5).

Short-cycle Plants

To produce liners for short-cycle forcing (Plate 7), cuttings are rooted in cell trays. The cell size depends upon the cultivar, the number of cuttings rooted in each cell, and the size liner desired. After the propagation phase, the newly rooted cuttings are moved into an optimum growing environment (see Chapter 5) and grown to the size needed by the forcer.

In addition to or instead of liners, some growers use a direct-stick program in which they propagate the cuttings in the actual pot that is used for forcing (Plate 6). This technique, developed by Moe (1973), is used mostly by large, specialized growers who produce year-round, short-cycle crops. This system is designed for 2.5 to 4.5-inch (6 to 11-cm) pot production and requires specific equipment and growing techniques not suitable for many growers. Table 3 describes such a system in use since 1988 at a major Danish pot rose producer (5 to 6 million units in 1992). A few large North American growers use a similar program adapted to their own growing facilities and geographical and climatic conditions. Direct-sticking can also be used to produce a prefinished plant for year-round finishing or to summer-propagate a long-cycle crop outdoors for production of dormant plants for winter forcing.

Long-cycle Plants

Long-cycle plants are grown under outdoor conditions during the summer and autumn by potting liners or direct-stick propagating in the desired final pot size to be used for winter forcing (Plate 8). If needed for moderating temperature, 20 to 35% shade can be used, but plants should be exposed to as much light as possible. Plants should be sheared periodically to encourage branching from the crown and in the basal part of the plant, and watered and fertilized as for a crop being forced (see Chapter 5). Chemicals should be applied to prevent black spot and powdery mildew (see Chapter 6). After the plants have gone dormant in the autumn, they must be protected from severe winter cold by covering or placing into cold storage (see Chapter 5).

Bare-root Rooted Cuttings

Similar to long-cycle plants, bare-root rooted cuttings are grown outside, except they are planted in media in raised benches. When the plants become dormant in the autumn, they are lifted from the benches, the media removed from the roots, and the plants placed into cold storage, where they are handled similarly to field-grown bare-root plants (see above). They are sold under the "X" grading system as described above for field-grown bare-root plants. A bare-root rooted cutting is the least common plant type used for finishing pot roses.

Plate 1. Examples of the range of flower color and form now available in pot roses (photograph by John W. Kelly).

Plate 2. One-year field production of T-bud grafted miniature rose cultivars to be used for forcing. The field is located in southern California (photograph by Jacques Ferare).

Plate 3. Orange Sunblaze grown as a patio tree (photograph by Jacques Ferare).

Plate 4. The stages of growth during forcing of a long-cycle crop with a cut-back plant on the left and a finished plant on the right (photograph by Jacques Ferare).

Plate 5. The well-developed root system of a long-cycle grown plant ready for forcing (photograph by H. Brent Pemberton).

Plate 6. An example of direct-stuck cuttings in the propagation phase (photograph by H. Brent Pemberton).

Plate 7. A well-rooted and vigorous liner ready for potting and forcing as a short-cycle crop (photograph by H. Brent Pemberton).

Plate 8. Plants being grown outdoors during the summer for a long-cycle production schedule (photograph by Jacques Ferare).

Plate 9. Long-cycle plants being grown in structures which can be covered for winter protection (photograph by Jacques Ferare).

Plate 10. Appearance of long-cycle plants shipped to a finisher and ready for forcing (photograph by H. Brent Pemberton).

Plate 11. Short-cycle plants: prior to the final pinch (right) and after the final pinch (left) (photograph by H. Brent Pemberton).

Plate 12. The plant on the right was grown under a spectral filter which increased the red:far-red light ratio in comparison to the one on the left (photograph by John W. Kelly).

Plate 13. Varying degrees of leaf chlorosis, showing a normal leaf on the left and one with severe chlorosis on the right. Iron deficiency or many kinds of root problems can cause this type of symptom (photograph by H. Brent Pemberton).

Plate 14. The response of Lady Sunblaze plants to uniconazole during summer forcing. From left to right are plants treated with 0, 25, 50, 75, and 100 ppm. Treatments were applied when new shoots were 2 inches (5 cm) long after the final pinch (photograph by H. Brent Pemberton).

Plate 15. Plants infected with *Cylindrocladium*. Note the collapsing stems (photograph by H. Brent Pemberton).

Plate 16. Close-up of plant infected with *Cylindrocladium*. Note the darkening and the fruiting of the fungus near the soil line (photograph by H. Brent Pemberton).

Plate 17. A close-up showing *Botrytis* infection on a flower and the subtending stem which developed after shipping (photograph by H. Brent Pemberton).

Plate 18. Orange Sunblaze flowers with normal flowers on the left and flowers on the right that are showing color and form abnormalities associated with high-temperature shipping (photograph by H. Brent Pemberton).

Plate 19. A properly handled plant on the left and a plant exposed to 82°F (28°C) during shipping on the right (photograph by H. Brent Pemberton).

Plate 20. Flower stages on Lady Sunblaze plants. From left to right, "pea-bud" and stages 1, 2, 3, 4, and 5. See also Table 11 for descriptions (photograph by H. Brent Pemberton).

Plate 21. Carts used for shipping finished plants in Europe (photograph by John W. Kelly).

Plate 22. Orange Sunblaze plants used in a patio container (photograph by Jacques Ferare).

5 Forcing

TIMING AND PLANT ESTABLISHMENT

Bare-root Plants

Because of the typically large size of field-grown, grafted, bare-root plants, they are usually placed into 7-inch (18-cm) or larger pots for forcing. This is the type of plant traditionally used for forcing larger flowering cultivars, but many miniature cultivars are produced this way as well (see Chapter 4). These plants are only available for a limited time during the winter and spring and are typically forced for Easter and Mother's Day, although techniques have also been developed for Valentine's Day forcing (see below). Bare-root rooted cuttings (usually miniature cultivars) are also available at this time of year but are usually potted at 1 per 4 to 5-inch (10 to 13-cm) pot for forcing.

Bare-root plants arrive dormant from the shipper. Dormant means that the plants are not in active growth; however, this does not mean that the plants are indestructible. Life processes are occurring that allow the plant to use stored carbohydrate reserves to stay alive. When exposed to the proper environment, growth will resume. If exposed to a stressful environment such as being allowed to dry out (see below), a dormant plant can be quickly lost. Therefore, shipping of plants to the forcer should be timed as close to the potting date as possible. The supplier has facilities for storing plants at the proper temperature and relative humidity and has some flexibility in shipping (see Chapter 4).

Bare-root plants should be kept in cold storage upon receipt until needed for potting. Short-term refrigerated storage of plants for a day or two in the plastic lined shipping boxes is acceptable provided that plants are tightly wrapped with the plastic liner to control moisture loss. If removed from the plastic liner, plants should be placed in cold storage and the humidity maintained as high as possible by spraying the plants with water several times daily. Regardless, plants should be checked for damage immediately upon receipt and claims filed if necessary. If the plants are frozen, leave in the plastic-lined boxes and store at 34–40°F (1–4.5°C) for 2 or 3 days to allow them to thaw gradually. Dormant rose plants can stand considerable cold if handled in this manner (MacKay, 1985).

Loss of moisture must be avoided when handling the plants during potting. Plants can be frequently sprayed with water or soaked for a few hours to maintain moisture content. Freshly harvested plants have a moisture content of about 48% and should not be allowed to fall below a level of 40% (Pemberton and Roberson, 1990). When freshly harvested field-grown plants were dried out for 24 hours at 60°F (15.5°C), moisture content fell to 33% and resulted in a 62% plant loss during subsequent forcing. Growth and flowering of surviving plants were delayed by up to 15 days compared to undried controls (Pemberton and Roberson, 1990). If plants have been exposed to excessive drying, soaking the roots in water for several hours will increase survival but will not hasten the forcing process already delayed by drying (Pemberton and Roberson, 1990).

Immediately prior to potting, canes on bare-root plants should be pruned to 6–8 inches (15–20 cm) from the graft union for polyantha or floribunda cultivars and to 4 inches (10 cm) for miniature cultivars. Proper pruning prior to potting is critical. Pruning too close to the graft union can weaken the plants and result in few flowering shoots. Pruning too high results in an unbalanced, top-heavy plant with only a few, dominant flowering shoots that force from the upper buds, leaving a bare base. Bare-root rooted cuttings should be pruned to 2 inches (5 cm) above the crown. Pruning to an outside bud is recommended for the polyantha and floribunda types because growth will be directed to the outside of the pot (Hammer, 1992). This is not necessary for the miniature types. For all types, weak and broken shoots should be removed. Roots should only be pruned enough to fit into the pot.

Grafted plants should be planted with the graft union just above the soil line, while bare-root rooted cuttings should be planted with the crown at or slightly below the soil line. After potting and placing into the forcing environment, a high-humidity atmosphere is needed to promote strong shoot growth (MacKay, 1985). Excessive water loss must be prevented until new roots begin to form and water can be taken up. However, care should be taken not to over-saturate the media. Plants can be sprayed frequently with a fog nozzle. Plants can also be covered with white opaque polyethylene or wet burlap, but covers should be removed when new leaves begin to expand (MacKay, 1985). Black or clear plastic should not be used because of the possibility of excessive heat buildup.

Forcing begins immediately after potting and generally takes 6 to 8 weeks. Top-grade plants (see Chapter 4) should be used because smaller sizes may require pinching, which delays finishing. All types of structures from cold frames to fully equipped, state-of-the-art greenhouses can be used for forcing; however, because temperature strongly influences forcing time (see below), the degree of control that is required must be considered.

Occasionally, a single dominant shoot will develop that should be completely removed at the base when three to four inches (7.5 to 10 cm) in length (Heins, 1981). If allowed to develop further, this type of shoot will suppress other flowering shoot growth and become disproportionately taller than the rest of the plant, thus resulting in problems with uniformity and shipping of the flowering plant. Pinching the rampant shoot is ineffective, as the resulting lateral shoots will also outgrow the rest of the plant.

Bare-root grafted plants being forced for Valentine's Day require special handling (Heins, 1981). These plants must be dug in late October and potted in November (Table 5). Availability can be a problem. Also, plants dug early may exhibit a certain degree of dormancy. Zieslin and Moe (1985) have proposed that buds on basal parts of rose plants require a period of cooling to be released from a partial state of dormancy. Two weeks of storage at 35–50°F (1.5–10°C) is adequate for uniform bud break during forcing and also shortens forcing time compared to non-stored plants (Asaoka and Heins, 1982). Storage for 4 weeks further shortens forcing time by as much as 1 week compared to 2 weeks of storage. Also, potting prior to storage for four weeks at 48–50° (9–10°C) shortens forcing time by about one week compared to plants stored bare-root and potted immediately before forcing (Asaoka and Heins, 1982). If storage facilities are available, the method of potting in early November prior to storage has the additional advantage of making use of the labor force available between Easter lily potting in October and poinsettia shipping in late November and early December (Heins, 1981).

Long-cycle Plants

Long-cycle plants are typically miniature cultivars that have been propagated and grown outdoors during the summer in the pot used for forcing (see Chapter 4). They go dormant in the autumn and will be leafless upon receipt. We have observed these plants to be difficult to force successfully in October, but they are usually ready to force by early December in most areas of North America.

As with bare-root plants, long-cycle plants develop a state of dormancy in the autumn and require a period of cooling for release from dormancy. Flower development was advanced by about 10 days and the number of flowering shoots per pot was increased when Orange Sunblaze plants grown outdoors in Pennsylvania were stored at 30°F (–1°C) for 16, 35, or 45 days beginning in early December (Zieslin and Tsujita, 1990b). Lady Sunblaze plants required 45 days of storage for the same effect.

Currently, dormant plants are stored outdoors until needed for forcing so that they receive the necessary cool-temperature exposure in the autumn and winter (Plate 9). During severe cold they can be protected with foam and plastic coverings as per standard practices for nursery stock. To avoid root damage, soil temperature should not fall below 25°F (–4°C). However, cold storage may have to be considered in milder climates for early forcing dates such as Christmas and Valentine's Day so that any cold requirement can be satisfied. Cold storage may also be needed for holding plants for forcing for Easter or Mother's Day crops when long-cycle plants are produced in mild climates, because forcing should not begin after natural bud break occurs in the spring, as it will cause reduced vigor and a longer forcing time after cut-back (Mor et al., 1986). Extended periods of cold storage should be avoided, however, as plants will begin to deteriorate when stored for over 12 weeks at 30–34°F (–1 to 1°C) (Smith and Skog, 1992).

Long-cycle plants should be unpacked immediately upon arrival and placed into the forcing greenhouse (Plate 10). If frozen, they should be treated as described above for bare-root plants. As the plants are unpacked, the shoots should be cut back to 2 to 3 inches (5 to 7.5 cm) above the pot and cleaned of all dead branches and debris. If potting into larger pots, repotting should be done as soon after cut-back as possible.

Long-cycle plants are typically grown in 4 or 4.5-inch (10 to 11-cm) pots (Plate 4). Forcing can be done in the same container, or plants can be potted into a larger size prior to forcing. Plants grown in 4-inch (10-cm) pots and potted into 6-inch (15-cm) pots at the start of forcing finish about the same time as those forced directly in the original pot but are larger because of the increased rooting volume. This allows a larger plant to be finished for Easter or Mother's Day than can be produced by using the short cycle with only 1 pinch (see below). However, plants potted from 4-inch (10-cm) pots to pots larger than 6 inches (15 cm) will probably need a pinch to produce a plant large enough. In this case, the use of a dormant bare-root plant could be a better choice.

Long-cycle plants are characterized by well-developed buds and root systems full of carbohydrate reserves (Plate 5); therefore, they can be forced under the lower light intensity of the winter months much better than a short-cycle plant (see below). For even growth, a good balance between light and temperature is needed (see below). Seven to 8 weeks are needed to finish long-cycle plants for Valentine's Day or 6 weeks if high intensity supplemental lighting is used (see below). Lighting can also be used for Easter or Mother's Day crops, but forcing should begin before natural bud break occurs in the spring, which can reduce vigor and result in a longer forcing time (Mor et al., 1986). Cold storage would therefore be necessary in mild climates to prevent bud break prior to forcing for a late spring crop.

Extremely vigorous, succulent, sucker-type growth will appear occasionally and should be completely removed. If allowed to continue growing, these sucker-type shoots will outgrow the rest of the plant, resulting in nonuniform flowering. As with bare-root plants, pinching this type of growth is ineffective.

Short-cycle Plants

Short-cycle plants are typically miniature cultivars that remain in active growth from propagation to finishing. The forcing phase begins with either rooted cuttings propagated in the final forcing container or liners (see Chapter 4). As a general rule, direct-sticking is used by large, specialist growers while liners allow more flexibility in choice of pot size and media mix (see below). The short-cycle method is used most effectively from March to November in North America, but can be successfully used from December to February with supplemental light and CO_2 (see below).

Large-scale producers usually direct-stick in 4-inch (10-cm) pots (Plate 6) in facilities equipped to produce a finished crop in the shortest time possible (see Chapter 4 and Table 3). By automating plant handling and controlling light intensity, photoperiod, CO_2, and temperature (see below) at each stage of production from start to finish at one location, a large number of homogeneous plants can be economically produced on a year-round basis if the market is available. Another use of a direct-stuck crop is as a prefinished plant for year-round finishing.

For producers who require a more heterogenous product mix or who have less sophisticated growing facilities, liners are probably more economical and allow more flexibility in the type of product finished. Liners usually consist of 1 to 4 cuttings rooted in each cell of a large cell tray and are shipped in active growth (Plate 7). They should be potted as soon as possible after receipt, but can be stored for up to 5 days at 35°F (1.5°C). These plants should never be exposed to freezing temperatures, as plants in active growth cannot withstand being frozen to the degree that dormant long-cycle or bare-root plants can.

One liner each (3 to 4 cuttings) is usually potted into 4 to 6-inch (10 to 15-cm) pots. Small liners (1 to 2 cuttings) are available for potting into 3-inch (7.5-cm) pots. Larger pots would require 2 or more liners or more than 1 pinch before finishing. Long-cycle or bare-root plants, therefore, may be better for producing larger pot sizes, especially for Easter and Mother's Day (see above). Another option would be to use a prefinished, direct-stuck, 4-inch (10-cm) grown plant as a liner for a larger pot size year-round.

The final forcing stage begins by pinching the plants to 1.5 or 2 inches (3.75 or 4 cm) above the pot rim (Plate 11). However, doing this without a well-established root system results in poor and uneven bud break, in turn leading to a finished plant with non-uniform flowering and poor balance. After direct-stuck cuttings have rooted (see Chapter 4), they must be allowed to develop until the root system fills the pot prior to pinching. Root systems must be similarly developed on freshly potted liners prior to pinching for final forcing. Two to 3 weeks are necessary to establish plants in 4-inch (10-cm) pots. A slightly longer time is needed for liners potted into 5 or 6-inch (13 or 15-cm) pots, but the finished plant will be correspondingly larger. Bottom heat is beneficial during this stage when greenhouse conditions are cool. Finishing time for a short-cycle plant is 5 to 8 weeks from the final pinch, depending on the time of year and local conditions.

THE FORCING ENVIRONMENT

Light

Light intensity is the most important factor affecting rose-plant growth and flowering. The correlation between cut-flower production and seasonal fluctuations in solar radiation is well known (Zieslin and Mor, 1990). A high incidence of blind (abortive) shoots on plants grown for winter cut-flower production has been attributed to low light intensity (Zieslin and Halevy, 1975; Zieslin et al., 1973). The benefits of supplementary lighting for winter cut-flower production have been reviewed by Tsujita (1987) and Zieslin and Mor (1990). Increased flower production in response to supplemental light has been attributed to increased bud break and enhanced mobilization of assimilates to the shoot tip, thus leading to a lower incidence of blind or non-flowering shoots (Zieslin and Mor, 1990). In addition to maintaining adequate light intensity, long photoperiods are also thought to be important for stimulating growth and flowering (Zieslin and Mor, 1990).

Supplemental lighting is now commonly used for winter production of pot plants (see Table 6 for light measurement conversions). Bare-root Garnette rose plants showed an increase in the number of flowering shoots per plant when they were forced under high-pressure sodium (HPS) light at 460 footcandles (61 µmol per second per square meter) from 5:00 A.M. to 5:00 P.M. in southern Michigan during winter (Asaoka and Heins, 1982). For short-cycle grown plants of Orange Sunblaze and Lady Sunblaze, Zieslin and Tsujita (1990a) demonstrated that flower-shoot formation increased with increasing light intensity up to 900 footcandles (119 µmol per second per square meter), the highest level tested, when continuous HPS lighting was used during September to December in southern Ontario. The maximum response was at 680 footcandles (90 µmol per second per square meter) during January to

February, however. Mortensen (1991) found the largest increase in flowering on short-cycle grown Orange Sunblaze plants during winter forcing in Norway when HPS light intensity was increased from 15 to 460 footcandles (2 to 61 µmol per second per square meter) for 18 hours per day; he found a smaller increase when the lighting was increased to 910 footcandles (120 µmol per second per square meter). Kyalo (1992) also found an increase in flower shoot formation in response to supplemental lighting during winter forcing as far south as northeast Texas (32°N latitude). Flowering was increased on Red Sunblaze and Orange Sunblaze plants with 600 footcandles (79 µmol per second per square meter) of HPS lighting for 8 hours at the end of the day. Lighting was also used during the day on cloudy days. Zieslin and Tsujita (1990a) found that a level of approximately 500 footcandles (66 µmol per second per square meter) of HPS lighting was needed simply to reach the light compensation point (the light energy level necessary for a plant to produce enough carbohydrates through photosyntheses during the day to compensate for the carbohydrates lost via dark respiration at night) at ambient CO_2 for plants of Orange Sunblaze and Lady Sunblaze. As a general recommendation, 16 hours per day of 500 footcandles (66 µmol per second per square meter) HPS lighting should be regarded as a minimum amount of light needed for successful winter production of pot roses in North America, particularly for Valentine's Day crops. For long-cycle or bare-root plants, 350 footcandles (46 µmol per second per square meter) may be adequate (Heins, 1981; McCann, 1991). A longer photoperiod of 18 to 20 hours is recommended by Mortensen (1991) at higher latitudes (approximately 60°N).

Floral initiation occurs very rapidly on new shoots forming after a pinch (Horridge and Cockshull, 1974). During these early stages of flower development, flower bud abortion can be promoted by low-light conditions (Moe, 1972; Moe and Kristoffersen, 1969). In southern Ontario, 1 week of 900 footcandle (119 µmol per second per square meter) HPS lighting at the beginning of a 5-week forcing period starting in mid-February increased flowering shoot formation on Orange Sunblaze plants as much as did lighting for the entire forcing period (Zieslin and Tsujita 1990b). A similar, but less pronounced response was seen with Lady Sunblaze plants. Lighting during early stages of forcing while plants are spaced pot-to-pot, followed by finishing at final spacing without lighting could represent a significant economic benefit over lighting during the entire forcing period. However, this technique should be tested on a small scale first, as cultivar and latitude could greatly influence its success.

The type of light source used is also critical to successful production. With cut roses, supplementary lighting with HPS or fluorescent lamps promoted bud break (Carpenter and Anderson, 1972; Khosh-Khui and George, 1977; Zieslin and Mor, 1990), whereas incandescent lighting reduced bud break (Moe, 1972). This has since been attributed to the ratio of the amount of red to far-red wavelength light energy omitted by the light source; a high ratio (HPS and fluorescent lamps) promotes sprouting and a low ratio (incandescent lamps) inhibits sprouting (Zieslin and Mor, 1990). Other effects of light quality have been observed in experiments with pot roses in which spectral filters were used. When grown under a filter that greatly reduced the far-red light component from natural light, Red Sunblaze plants were found to be shorter and darker green than plants under light filtered through water only, though the number of flowering shoots was not affected (Plate 12) (McMahon and Kelly, 1990). Consequently, incandescent lighting, which has a low red to far-red light ratio, should never be used for pot rose production.

High-pressure sodium vapor lamps are the most commonly used source of supplemental lighting because of their high efficiency for converting electrical power to radiant energy (Tsujita, 1987). When light fixtures are installed, units should be arranged according to the manufacturer's specifications. Most firms will provide assistance to determine the best installation design for achieving the desired level of lighting for any given greenhouse situation. Light-energy levels decrease by the square of the distance above the plant canopy. In other words, the light intensity is reduced by a factor of 4 if the height of the light source above the plant canopy is doubled. Also, a certain amount of light-pattern overlap is necessary to achieve maximum uniformity. Fixtures designed for greenhouse use are compact for minimal shading and have reflectors that provide the lighting pattern needed; 400-watt (W) fixtures are most commonly used in greenhouses with limited height, whereas 1000-W fixtures can provide the uniformity and intensity needed with fewer fixtures in a high-profile house.

Temperature

Pot roses are commonly grown with a 62 to 65°F (16.5 to 18°C) night temperature and a 68 to 70°F (20 to 21°C) day temperature (McCann, 1991). A constant temperature between 64 and 68°F (18 and 20°C) can be used in greenhouses with a high degree of climate control. However, night temperatures should not be higher than day temperatures. Drastic changes in temperature over a short period should be avoided as malformed flowers could result. For example, if the night temperature is raised by a few degrees to speed cropping time, it should be done by 1 to 2°F (0.5 to 1°C) per night instead of all at once.

Not surprisingly, time to flowering is directly related to the forcing temperature (Kyalo et al., 1996; Moe, 1973; Mortensen, 1991). Increasing the forcing temperature also reduces plant height (Kyalo et al., 1996; Moe, 1970; Mortensen,1991), thus potentially reducing the need of a growth retardant for height control. However, extremely high temperatures should be avoided. Constant forcing temperatures above 81°F (27°C) resulted in pale flowers on Orange Sunblaze plants (Mortensen, 1991). Net photosynthesis (Pn) decreased at temperatures over 75°F (24°C) for plants of Red Rosamini, Orange Sunblaze and Lady Sunblaze (Jiao et al., 1990) but remained high for plants of Sunbird even at temperatures over 90°F (32°C). Cultivar selections should be made carefully for attempting summer production in hot summer areas. Shade and fan and pad cooling systems are commonly used to control greenhouse day temperature in the summer. Twenty to 35% shading can be used; 30% is common. Shade greater than 35% should not be used even under high light and temperature conditions. Many cultivars such as Red Sunblaze and Orange Sunblaze will grow acceptably as long as day temperatures are kept under 90°F (32°C).

Time to flower is also related to bud diameter (Asaoka and Heins, 1982). Table 7 shows days to flower for various bud diameters at 61°F (16°C) night temperature and 68°F (20°C) day temperature. Bud diameter can be measured with an engineer's template for drawing circles (Heins, 1981). Higher temperatures presumably hasten flowering whereas lower temperatures, especially below 55°F (13°C), delay flowering (Heins, 1981). This procedure has not been extended for use with other temperatures or cultivars but could be used as a guide for developing such information.

For winter and spring forcing of bare-root and long-cycle plants, some sources recommend that plants be kept cool (45–50°F; 7–10°C) when starting to force (MacKay, 1985; Laurie et al., 1969), but starting at a 60–62°F (15.5–16.5°C) night temperature has been found to result in a more uniform and well-proportioned plant (Hammer, 1980; Moe, 1970). Maintaining a balance between light and temperature is important, however. A night temperature of 65°F (18°C) resulted in excessive flower bud abortion when bare-root plants were forced under natural light for Valentine's Day in Michigan (Heins, 1981). A night temperature of 60 to 62°F (15.5–16.5°C) or supplemental lighting (see above) alleviated the problem. A natural progression of increasing light intensity and temperature exists outdoors during mid-spring in most of North America, thus making forcing outdoors or in cold frames a viable opportunity for late spring sales.

Carbon Dioxide

Carbon dioxide (CO_2) enrichment is most commonly practiced during the winter months at latitudes greater than 40°N or anywhere infrequent ventilation allows economical maintenance of a 1000-ppm concentration (Hicklenton, 1988). The development of new greenhouse-control systems and crop-production methods is making CO_2 enrichment economically attractive at southern latitudes as well (Hicklenton, 1988). However, the benefits of using CO_2 enrichment for pot rose production have not been as clear-cut as those seen for cut-rose production. The responsiveness of cultivars must be assessed when CO_2 enrichment is used. No increase in flowering was found for Orange Sunblaze plants exposed to as much as 1050 ppm CO_2 during winter production (Clark et al., 1993; Mortensen, 1991). However, Lady Sunblaze plants had 42% more flowers than those grown under ambient CO_2 conditions when exposed to 800 ppm CO_2 regardless of supplemental HPS lighting from 300 to 1200 footcandles (40 to 158 µmol per second per square meter) (R. Moe, personal communication, 1988).

The method of applying CO_2 can also affect the flowering response of pot roses. When grown under diurnally changing CO_2 treatments (increasing or decreasing between 600 and 1500 ppm in 4 steps of 300 ppm during the natural light period plus holding at 900 ppm CO_2 during a night-time supplemental light period—total photoperiod of 18 hours) or constant 600 ppm (18 hours per day), Red Minimo plants had more flower buds than those held under constant 900 ppm (18 hours per day) conditions, although the number of shoots was the same (Andersson, 1991).

CO_2 enrichment generally results in increased stem length for rose cut-flower production (Mastalerz, 1987). This is not necessarily an advantage in pot rose production. Orange Sunblaze plants were taller when grown in 700 ppm CO_2 under ambient winter light conditions in South Carolina (Clark et al., 1993). Lady Sunblaze were also taller when grown in 800 ppm CO_2 under 300 footcandles (40 µmol per second per square meter) HPS light for 18 hours per day during winter in Norway (R. Moe, personal communication, 1988). However, when supplemental light levels of 600 to 1200 footcandles (79 to 158 µmol per second per square meter) were used, CO_2 levels had very little effect on height.

No hastening of flowering was seen when 800 to 900 ppm CO_2 were used with 300 to 1200 footcandles (40 to 158 µmol per second per square meter) of HPS supplemental lighting during winter forcing of Orange Sunblaze or Lady Sunblaze plants in Norway (R. Moe, personal communication, 1988; Mortensen, 1991). However, Orange Sunblaze plants treated with 700 ppm CO_2 flowered 6 days earlier than those under 350 ppm when forced under ambient winter light conditions in South Carolina (Clark et al., 1993).

Many variables are involved, therefore, in determining the effectiveness of CO_2 enrichment for pot rose production. Jiao et al. (1990) determined that at saturating irradiance and 72°F (22°C), net photosynthesis in plants of 4 miniature pot rose cultivars was saturated at 800 ppm CO_2. A level of 800 to 1000 ppm CO_2 is generally recommended, especially for winter production of short-cycle grown plants (McCann, 1991). However, the effect of CO_2 enrichment on flowering and final plant form should be assessed under local conditions to determine whether this practice is needed for enhancing plant quality.

Media and Watering

Pot roses should be grown in a well-drained media with good water-holding capacity. Any peat-lite mix such as 1 peat: 1 perlite: 1 vermiculite by volume is desirable. No differences were seen when Red Sunblaze short-cycle plants were grown in 17 different commercially available potting media during the winter in northeast Texas (B. Pemberton, G. McDonald, and W. Pianta, 1992 unpublished results). Optimum pH for pot rose production is 5.5 to 6.0; readings above 7.0 result in chlorosis. Incorporating superphosphate and dolomitic limestone in the mix helps balance the pH and provides calcium and magnesium.

Either too much or too little moisture can result in leaf yellowing and abscission and root loss on rose plants (White and Holcomb, 1987). Overwatering is especially harmful for dormant long-cycle plants at the beginning of forcing into growth, for newly potted liners (long- or short-cycle plants), and for plants in all types of production schedules that have just been pinched or cut-back while in active growth. Overwatering can also be critical for newly establishing bare-root plants, but high humidity must be maintained around the canes without saturating the media until the leaves begin unfolding.

On the other hand, allowing plants to dry out only once can also ruin a crop. Roses do not recover well from a missed watering, so careful monitoring is critical. Excessive drying of the media can also result in soluble salt injury if fertility levels are high (White and Holcomb, 1987). Drought stress becomes especially critical as flowers reach pea-bud size. At this stage, overwatering is hard to do at normal forcing temperatures.

Water quality should also be considered for successful pot rose production. Water supply pH and soluble salt content can vary widely across North America, even within individual states or provinces. Water to be used for irrigation should, therefore, be tested on a regular basis. Instructions for sending water samples can be obtained from your local Cooperative Extension Service or from a commercial lab. Soluble salt levels are measured with a conductivity meter, which measures electrical conductivity (EC). EC readings are expressed in mhos/cm and will be reported as EC $\times 10^{-3}$ (millimhos per cm), EC $\times 10^{-5}$, or EC $\times 10^{-6}$ (micromhos per cm). These units are easily convertible from one to another. Measurements of individual ions will be reported in either parts per million (ppm), milliequivalents per liter (meq per liter; see White, 1987, for explanation), or both.

Total soluble salts above 910 ppm (1300 micromhos per cm) and a bicarbonate ion (HCO_3^-) level of 112 ppm in the irrigation water reduced the yield of cut-flower roses (Hughes and Hanan, 1978;

White and Holcomb, 1987). An excess of these levels should be avoided for pot rose production. The bicarbonate ion is highly toxic to roses (Hughes and Hanan, 1978) and should be kept below 90 ppm (1.5 meq per liter). Sodium and chloride ions can also be harmful (Hughes and Hanan, 1978) and should be maintained below 46 ppm (2 meq per liter) and 70 ppm (2 meq per liter), respectively.

Nutrition

Much is known about rose-plant nutrition because of the long history of cut-flower production and research. Nutrient imbalances, deficiencies, and toxicities can be serious because cut-rose plants are kept in cultivation for several years. However, compared to cut-flower production, pot rose nutrition is simplified considerably because of the relatively short production schedules.

A commonly used feeding program consists of 180 to 200 ppm nitrogen with a 2-feed, 1-leach schedule and a reduction of nitrogen upon finishing (McCann, 1991). Any good, commercially available fertilizer such as 20–20–20 or 15–10–12 or equivalent can be used. Supplementing a constant feed schedule with a topical application of a balanced, slow-release fertilizer can be beneficial for maintaining nutrient levels. A weekly liquid application of iron chelate and magnesium at 5 and 10 ppm, respectively, can be used to ensure good foliage color (especially the iron). Micronutrients should be incorporated in the feeding program to ensure proper growth.

Routine monitoring of media pH and soluble salt levels in conjunction with visual symptoms is a simple, effective tool for determining whether nutritional problems exist and whether further sampling is needed. Every producer should have pH and conductivity meters and know how to use them properly, as these measurements can indicate a problem before the visual symptoms occur. Common visual symptoms are described in Table 8. If problems are suspected, detailed media and foliar analyses done on a regular basis are helpful for monitoring and solving nutritional problems.

Instructions for sending media and foliar samples can be obtained from your local Cooperative Extension Service or from a commercial lab. Specifically, media samples should be taken as core samples from the full profile in the pot, and media from several pots should be combined for a composite sample. The newest, undamaged, fully expanded five-leaflet leaves should be sampled for foliar analysis. Leaves from several shoots on several plants will be needed to make up one sample. For both types of samples, separating samples from different parts of a large production range or by cultivar is more informative than a composite sample from thousands of plants.

A full analysis should be requested for both media and leaf samples. Table 9 has values that can be used for interpreting results from media testing performed at university or commercial labs. Also, testing labs usually furnish values that can be used for interpretation. However, interpretation must take into account the analysis method used. The biggest difference among labs in the analysis methods used for different media types is in the extraction technique (Mastalerz, 1977; Warncke and Krauskopf, 1983). For artificial media mixes, some type of water extraction method is commonly used. The ratio of water to soil can vary from a saturated paste (referred to as the Saturated Media Extract method or SME) to 5:1. Some labs use a mild acid extraction for artificial media, referred to as the Spurway method. A large commercial testing firm recommends the SME for media mixes with less than 30% natural soil. Since the majority of pot roses are grown in an artificial mix, the SME or some other type of water-based extraction method will most likely be used for media analysis. The same firm recommends a method based upon an acid extraction solution for media mixes if the natural soil component is greater than 30%. This method is commonly referred to as Morgan or modified Morgan. It is important to remember that producers must use the appropriate values to interpret the media test. The commercial or university lab that did the testing will be the best source of critical values to interpret the results. Also, do not expect to get the same values for the same media sample sent to different labs. Developing a history with a lab you feel comfortable with is recommended.

Interpretation of leaf sample analyses is more straightforward than for media samples. Table 10 contains critical values for interpretative use. These values have been arrived at primarily through studies of cut-flower varieties, but they form a good basis for determining nutritional disorders for all types of roses. When used in conjunction with visual symptoms (Table 8), these values can be helpful for explaining a nutritional disorder so that it can be corrected. If a more detailed treatment of nutritional disorders is desired, refer to White (1987).

Nutritional demand can vary depending on climatic, environmental, and geographical conditions. For example, plants grown under high levels of supplementary light or in a geographic location with a warm, high-light climate may need more fertilization than plants under ambient light conditions or a cool, cloudy climate, respectively. Plants suddenly exposed to sunny, warm conditions after being held in shady, cool areas can exhibit symptoms of micronutrient deficiency (e.g. copper) in the young foliage. This is usually an uptake problem which can occur when the warm conditions result in a high shoot demand for nutrients combined with the lack of root activity in the still-cold media that has not had enough time to warm due to the suddenness of the environmental change. These conditions occur most frequently during late winter and spring during forcing in cold frames or structures without cooling systems.

Fertilization programs should also be tailored to the type of plant being used for forcing. Fertilization of short-cycle plants should be maintained continuously so that no checks in growth occur during production. Long-cycle plants usually exhibit a loss in root activity during storage. This can manifest itself as an interveinal chlorosis visible on new foliage at the start of the forcing period (Plate 13). It is usually a transient problem, but care must be taken not to overwater or fertilize excessively until root activity is restored. If the chlorosis persists, other problems, such as a root rot, could be present. To prevent burning of the new roots, bareroot plants should not be fertilized until active shoot growth begins (MacKay, 1985).

Pot Size and Spacing

The choice of pot size depends upon many factors such as the type of plant used for forcing (see Timing and Plant Establishment section above), postharvest considerations (see Chapter 7), and marketing strategy (see Chapter 9). An additional consideration is that of aesthetics. Sachs et al. (1976) suggested that the ratio of the height of the plant and the pot to the width of the plant should be 1.5 to 1.7. Many of the new miniature cultivars grown in 4 to 5-inch (10 to 12.5-cm) pots fit this criterion without the use of a growth retardant (Mor et al., 1986). Even with a wise combination of cultivar and pot size, however, a growth regulator may be useful for producing an aesthetically pleasing product (see below), especially under certain environmental conditions.

Proper spacing allows uniform canopy development and prevents the flowering shoots from becoming entangled during production, thus reducing damage during bench removal for shipping. It also promotes air movement, which aids in the control of certain diseases and pests such as powdery mildew, downy mildew, and spider mites. Spacing can be pot to pot until 2 weeks after the final pinch (McCann, 1991). Final spacing for miniatures should be 3 to 4 pots per square foot (32.5 to 43 pots per square meter) for 4-inch (10-cm) pots, 2.5 to 3 pots per square foot (27 to 32.5 pots per square meter) for 4.5-inch (11-cm) pots, or 1 to 1.5 pots per square foot (11 to 16 pots per square meter) for 5 to 6.5-inch (12.5 to 15.5-cm) pots. Bare-root plants forced in 6-inch (15-cm) pots should be placed on 10 or 12-inch (25 or 30-cm) centers, which is 1 to 1.4 pots per square foot (11 to 15 pots per square meter) (Heins, 1981). Larger pot sizes should be spaced proportionally farther apart.

Growth Regulators

Plant height at flowering is a result of many factors, including light, temperature, CO_2, cultivar, and type of plant used for forcing. For example, many of the newer miniature cultivars grow more in proportion to finished pot size than the older polyantha and floribunda types without a growth regulator (Mor et al., 1986). Also, rose-plant response to growth regulators can be dependent on several factors (see below). As with all chemicals, applications on small groups of plants under local conditions should be tested before large-scale applications are made.

Moe (1970) found chlormequat (Cycocel; CCC) to effectively control the height of Garnette, Margo Koster, and Morsdag roses grown from bare-root plants. The best response was obtained by spraying to runoff with 1000 to 2000 ppm active ingredient (a.i.) when new growth was 2 to 2.5 inches (5 to 6.5 cm) in length and again 8 to 10 days later. Sprays over 2000 ppm caused severe foliar damage. Also, CCC improved the flower color and darkened the foliage, but it reduced the number of lateral flowers on the flowering shoots and delayed flowering by about 4 days.

Drench applications of CCC have produced variable results. When applied to Garnette and Margo Koster plants in a peat-perlite media, 1 g a.i. CCC (i.e. 100 ml of a 1.0% a.i. solution) per 6-inch (15-cm) pot resulted in growth control without foliar damage (Moe, 1970), but when applied to Morsdag plants in a peat-clay media, there was no response to the same rate. A 2-g a.i. solution resulted in severe leaf injury when applied to Garnette and Margo Koster plants. However, when Mor et al. (1986) made 2 applications 2 weeks apart of 3 g or 9 g a.i. CCC per 4-inch (10-cm) pot to American Independence long-cycle plants beginning at cut-back, growth was unaffected and no damage was reported, but the number of flowers per plant was greatly increased. Promotion of flowering by drench and spray applications of CCC has been observed in other types of roses (Mor and Zieslin, 1987), but the discrepancies between the various studies are not understood.

Moe (1970) also found that daminozide (B-Nine) effectively controlled the height of Garnette and Margo Koster roses grown from bare-root plants. Foliar sprays to new growth 2 to 2.5 inches (5 to 6.5 cm) long of up to 500 ppm reduced growth. However, sprays of 1000 ppm reduced growth without foliar damage, but resulted in pale flower color, especially on plants of Margo Koster. Despite these results, B-Nine has not been observed to be very effective on miniature pot roses, a subject for which research results are lacking.

Ancymidol was more effective for height control as a soil drench (0.25 mg a.i. or 0.5 mg a.i. per 4-inch or 10-cm pot) than as a foliar spray (100 ppm) when applied twice (10 and 20 days after cut-back) to American Independence long-cycle plants (Mor et al., 1986), but no effect of sprays or drenches was found on growth of Garnette plants in the same study.

Uniconazole (Sumagic), a triazole, is a relatively new growth regulator which has activity on roses. When sprayed with 20 or 30 ppm 3 weeks after potting, height of long-cycle Orange Sunblaze plants was reduced by 32 and 39% at flowering, respectively (Pobudkiewicz and Goldsberry, 1989). Height reduction averaged 26% when short-cycle plants of Red Sunblaze, Orange Sunblaze, and Lady Sunblaze (Plate 14) were sprayed with 25 ppm Sumagic when shoots were 2 inches (5 cm) long after pinching (Kyalo et al., 1993). In both studies, flower numbers were unaffected, peduncles were shorter, the leaves were darker green, and flowering was delayed by about 3 days on treated plants. Higher concentrations severely reduced plant height, number of flowers, and peduncle length, and caused a longer delay in flowering (Kyalo et al., 1993). Also, application of Sumagic to new shoots that had grown 2 inches (5 cm) after pinching was as effective in reducing plant height as was a split application (half the full rate at each time) made when new shoots were 2 inches (5 cm) long and when new shoots had pea-size flower buds (Kyalo et al., 1993). A single application at pea-bud stage was not effective in controlling height. A single spray of 20 to 25 ppm applied before flowers buds are visible appears to be the most effective treatment. However, care should be taken to test concentrations under local conditions, as this chemical can be more active under the low-ambient-light conditions of winter and early spring than during summer (Kyalo et al., 1993).

Though few studies have been published concerning efficacy, paclobutrazol (Bonzi) has been observed to be effective on pot roses. Two sprays of 50 ppm or 3 sprays of 33 ppm applied 7–10 days apart have been found to reduce plant height. Sprays should be applied before buds are visible for maximum effectiveness.

Because of variable results, CCC, ancymidol, and B-Nine are not commonly used for pot rose height control. Bonzi and Sumagic show promise as effective growth regulators for pot roses. However, local trials are strongly recommended to determine the optimum rates under local conditions. As with all chemicals, the label should be consulted before use and instructions closely followed. Changes in the label can occur rapidly in response to experimental results.

6 Pests and Diseases

Although roses are relatively easy to grow, the crop should be inspected frequently to minimize damage from insects and diseases. Tremendous advances have been made in recent years that reduce the greenhouse grower's dependence on scheduled chemical applications. By following the principle of integrated pest management (IPM), growers increase the frequency of careful crop inspection and use chemical control only when needed. With IPM every effort is made to use cultivars less susceptible to insects and diseases, to provide optimal cultural and environmental management of greenhouse crops, and to develop barriers and traps to reduce infestations. Careful monitoring for insects and diseases will allow the grower to make spray applications only when needed, thereby reducing labor and chemical costs as well as environmental risk. Even when following careful management practices, many growers find it necessary to spray for the 3 major problems of greenhouse pot rose production (aphids, spider mites, and powdery mildew) every 7 to 14 days. Chemical recommendations are not included in this book because labels and restrictions change quite frequently.

Because some pesticides are phytotoxic to roses, it is advisable to always treat a small section of plants and wait for 3 to 4 days to determine damage potential before using a new pesticide. Always carefully follow manufacturer labels on pesticides to comply with legal requirements and minimize the chance for phytotoxicity. Growers have reported that pesticides applied when roses are under water stress increase the incidence of leaf drop (Miller, 1987).

INSECT AND MITE MANAGEMENT

Forced roses are usually sold after 6–17 weeks in the greenhouse and are, therefore, not subject to as many insect and mite problems as longer-grown pot crops. However, there are some pests that can be bothersome for the greenhouse grower. The most common insect and mite pests are aphids, thrips, and spider mites. Identification and control of insects is made easier with the knowledge of the type of mouthparts on a particular species and the symptoms of damage from the different types of mouthparts. Two types of mouthparts are prevalent in pests of potted roses—rasping-sucking and piercing-sucking.

Insects such as aphids pierce the foliage with structures similar to needles called stylets. After the foliage has been pierced, the insect siphons the plant fluids through the stylet and into the insect's digestive tract. Symptoms of this type of damage are chlorotic areas on the foliage, curling and deformed new growth, and, occasionally, wilting of tissue. Stomach poisons that are sprayed on the surface of the foliage are not effective on this groups of insects because the insects do not consume the foliage, only the foliar fluids. Systemic insecticides, which are carried through the intercellular fluids, are most effective. Contact poisons are also effective insecticides for this class of insects, since the residue is located where the insect will make contact with it in its feeding. Spider mites also fall into this group of pot rose pests.

The 2-spotted spider mite is a serious pest associated with rose production. Spider mites on roses can be difficult to combat because of rather dense foliage. It is, therefore, critical to get good spray penetration into the foliage and to contact the underside of the leaf with the spray, as this is where spider mites tend to reside. Mite problems increase substantially as greenhouse temperatures increase and during periods of lower humidity.

Thrips is the most common insect pest with a rasping-sucking type of mouthpart found in rose production. They feed primarily on the rose flower but will also cause foliar damage. Thrips stylets are used as cutting or rasping devices to tear the petal or foliage surface open, so that the insect can feed on the intercellular fluids. Damaged plant parts appear discolored, mottled, or streaked. New growth is often distorted, and buds fail to open. Systemic insecticides and contact poisons are effective on this pest group. Screening the greenhouse with fine mesh materials to reduce the entry of thrips and other insects will reduce the number of spray applications required.

DISEASE MANAGEMENT

Disease organisms may be introduced into the greenhouse by insect vectors or worker handling in addition to airborne and water-splashed transmittal from diseased plants brought into the greenhouse. It is critical that good sanitation be practiced at all stages of greenhouse production and through shipping and storage. Plant diseases can be minimized by carefully selecting cultivars, controlling humidity, using horizontal air flow fans, removing infected tissue from the greenhouse, and avoiding splashing water on plant tissue.

Several different types of disorders affect pot roses. One group of diseases are the "rots." Root rots cause a rotting of the roots from the ground line and below or from the crown, and roots of plants become soft and mushy. Stem rots are manifested as stem and whole-plant decay at or above the soil level, from which the stem or plant collapses and dies.

Cylindrocladium scoparium causes black, water-soaked cracking of the bark near the soil line in potted roses (Plates 15 and 16). This fungal stem rot has become a significant problem as the production of pot roses has increased in North America. It is most damaging during propagation, where its action is cyclical. For example, it is most active in Pennsylvania in late summer and early fall, whereas in California it mostly affects crops propagated from May to September. However, the fungus is always present, but seems to be more active under warm, humid, low light conditions and when plants are under stress. Very little is known about the life cycle of *Cylindrocladium* and about the conditions that affect its activity on roses, so disease control is difficult. The best control of *Cylindrocladium* is the implementation of sound sanitary practices, including the purchase of only clean stock and the prompt removal and destruction of diseased plants.

The most heavily publicized and frequently epidemic rose disease is known as black spot. The symptoms of this fungal disease are circular or irregular areas on leaves which first show yellowing around the border and then develop a black lesion. Heavy defoliation usually occurs. Plant breeders have worked for years to develop rose varieties which are resistant to black spot. They have recently been successful in finding some genetic resistance, although the level of resistance seems to vary significantly according to the region of the country where roses are grown. Black spot is commonly found in the landscape particularly when overhead irrigation is used. Producers who use outdoor shade-house or open-field production of potted roses will need to make frequent fungicide applications to control black spot due to prolonged periods of wet foliage under outdoor conditions. Fortunately, black spot rarely occurs during greenhouse production.

Powdery mildew is one of the most common fungal diseases encountered in the production of pot roses. Symptoms include a white, powdery substance that may develop on all plant parts, including upper and lower surfaces of leaves. Young leaves are most susceptible and may become twisted and distorted when infected. Under severe infection, shoot tips may be killed and flower buds may fail to open. There are effective chemicals on the market for control of powdery mildew. The critical factor for chemical control is to begin fungicide applications as soon as weather conditions are conducive and symptoms occur. Environmental controls such as reducing night greenhouse humidity through adequate ventilation is one sound approach to discourage development of powdery mildew. Continuous air movement from horizontal air flow fans will also help prevent powdery mildew. Although rose varieties are being developed that exhibit resistance to powdery mildew, the resistance is difficult to retain because new races of the fungus develop that overcome the resistance (Horst, 1983).

Botrytis blight is a common pot rose fungal disease. It is frequently a problem during shipping and storage of plant materials (Plate 17), but it also occurs during production. Dormant roses held in storage may develop a fuzzy growth; the young canes are the most susceptible, and portions of the plant are killed. *Botrytis* blight usually occurs under humid conditions; the gray mold of the fungus becomes visible and leaves become blighted and turn dark-brown to grayish in color. Controlling relative humidity levels and dampness in the greenhouse helps reduce the incidence of

Botrytis. The disease often damages flowers, resulting in buds that fail to open or open petals that develop a grayish cast and premature death of the flower. All Botrytis-infected tissue should be removed from the plant and carried outside the greenhouse. Condensation and high humidity can be reduced by ventilation of the greenhouse. Chemical controls are available for *Botrytis*, but resistance to fungicides can become a problem.

Another fungal disease that can cause severe losses in pot rose production is downy mildew. Symptoms include purplish red to dark brown, irregular to angular spots, and severe leaf abscission. Under cool, humid conditions this disease can strike suddenly and spread very quickly. Maintaining humidity below 85% through ventilation can prevent the development of the disease. It is most common in unheated plastic-covered structures used for spring forcing, but can appear and require preventative sprays whenever humidity and temperature conditions are ideal.

An excellent reference on rose diseases is available (Horst, 1983) that details diseases, organism life cycles, and control strategies.

7 Postharvest Handling

The grower's responsibility for a crop does not end with the removal of the plants from the greenhouse bench. The weeks of effort to produce a high quality crop may be wasted if the crop is not properly transported to market and, subsequently, into the hands of the consumer. Postharvest handling includes all the steps to move the product from producer to consumer and is the phase most often neglected in the production and distribution of floricultural crops. Approximately 20% of all losses in floricultural crops have been attributed to postharvest losses (Staby et al., 1976). These losses directly affect grower and retailer profits and, more importantly, consumer attitudes toward flowering potted plants. Each time someone fails to carry out proper postharvest handling practices conscientiously, quality is irreversibly lost and consumer confidence needlessly sacrificed.

Typical problems that are encountered during the postharvest life of miniature potted roses include flower color and form abnormalities (Plate 18), flower abscission, leaf yellowing and abscission (Plate 19), and the formation of etiolated lateral shoots during dark shipping. Many factors strongly influence the appearance and severity of these phenomena, including genetics, production techniques, stage of harvest, handling techniques, sanitation, storage temperature and duration, the presence of ethylene, and the retail and consumer environments. Proper management of each of these factors helps to ensure that a high-quality potted rose is received and used properly by consumers. This, in turn, encourages repeat sales.

FACTORS DURING PRODUCTION THAT AFFECT POSTHARVEST LIFE

Postharvest quality of all floricultural crops is influenced by grower practices during production (Halevy and Kofranek, 1976; Shafer, 1985). Watering frequency, fertility regimes, CO_2 enrichment, season of year, temperature, light, sanitation, and cultivar selection are all critical production factors that may significantly alter postharvest performance of potted crops.

Irrigation

Potted roses are highly susceptible to water stress during shipping and handling and in the hands of consumers. To promote good postharvest quality, irrigation during production should be carried out in a way that allows the plant to develop an extensive root system free from disease. Work with cut roses (White, 1987) showed that allowing the media to dry out excessively even once can cause older leaves to yellow and abscise. Plants should be thoroughly irrigated 12–18 hours without fertilizer before shipping to allow for sufficient drainage to occur so that the shipping carton does not become excessively moist.

Nutrition

Improper fertilization during production can lead to major problems during postharvest handling. Soluble salts in the media should be monitored during production. Excess soluble salts (see Chapter 5) should be leached from the media so that foliar injury does not occur during postharvest handling. The concentration of soluble salts from fertilizers increases significantly if the growing medium is allowed to dry out. Soilless media are quite susceptible to drying out during prolonged storage and in the retail environment. If the medium dries excessively, injury to the leaf tips is likely to occur; leaf abscission will result from severe drying.

High production nitrogen rates may lead to weak, succulent, and spindly growth that is easily damaged during handling and shipping. Excess nutrient levels in the plant tissue may also contribute to decreased flower and foliage longevity and increased incidence of disease. Plants grown with too little fertilizer often have weak stems and exhibit leaf chlorosis and abscission during shipping and handling. Reduction in fertilization frequency and nitrogen (N) concentration has been shown to be beneficial for postharvest longevity of some crops. However, constant feeding of up to 300 ppm N has been shown to be nondetrimental to pot rose longevity (Clark, 1990). In general, potted rose plants can be constant-fed with 200–300 ppm N from a complete balanced fertilizer to prevent fertility-related postharvest disorders (Clark, 1990). Although terminating fertilization prior to shipping has not been shown to be necessary with pot roses, a thorough watering without fertilizer that results in a good leaching prior to sleeving and shipping is always a good idea. See Chapter 5 for further details on fertilization.

CO_2 Enrichment

CO_2 enrichment is often used during potted rose production to increase plant growth and reduce cropping time (see chapter 5). In a study to evaluate the influence of production CO_2 on postharvest characteristics of potted roses, rooted liners of potted roses were transplanted into 4-inch (10-cm) pots and grown as short-cycle plants during two seasons in a greenhouse enriched with CO_2 at 350, 700, and 1050 ppm under standard production practices (Clark et al., 1993). After 5 days of storage, plants were placed in a lighted interior environment for postharvest evaluation. Immediately after removal from storage at 40°F (4.5°C), plants that had been grown in 350 ppm CO_2 had fewer etiolated shoots than plants grown at 1050 ppm. Storage at 61°F (16°C) intensified this effect. However, CO_2 concentration during production was not related to postharvest leaf chlorosis. The primary benefit of using CO_2 enrichment with pot roses is to increase the production efficiency of the crop (see Chapter 5). If the grower uses CO_2 enrichment, however, the crops must be shipped at low temperatures or for very short durations to prevent etiolated shoot development during storage.

Season of Production

The time of year that a crop is grown may affect the longevity of the crop once it is placed in an interior environment. Experiments in Texas under glasshouse and growth-chamber conditions with short-cycle plants have shown that poststorage floral longevity is longer for Orange Sunblaze and Red Sunblaze plants forced under summer-like conditions than for those forced under winter-like conditions (Chen, 1990; Kyalo, 1992; Kyalo et al., 1996). Flower color and form abnormalities that appeared during the interior evaluation after storage were worse for plants forced during the winter than for those forced during the summer (Chen, 1990). In Denmark, postharvest longevity was greater for plants of Dreaming Parade, Victory Parade, Femini Rosamini and Red Minimo grown during April to June than those grown from January to March or August to October (A. S. Andersen, M. Serek, and P. Johansen, personal communication, 1994). In addition, long-cycle plants forced under natural light and flowered in January/February in northern Germany had a lower shelf-life than plants of the same variety forced in April/May (L. Hendriks, personal communication, 1988). In April/May, the flowers lasted longer and the tight buds opened, whereas in February, the few buds on the plants dried out and did not open fully. Seasonal differences in postharvest longevity are probably due to variation in light intensity and day length between the growing seasons, although differences in forcing temperature may also be important. Increased levels of supplemental lighting and warmer growing temperatures were beneficial to postharvest longevity of cut flowers compared to a cooler growing environment (Fjeld et al., 1994; Gudin, 1992; Moe, 1975). Little is known about how the individual factors of supplemental lighting used commonly for winter production and the forcing temperature affect postharvest longevity of pot roses. Silver thiosulfate and benzyladenine have been used to prolong postharvest longevity of pot roses in general (see section on Chemical Treatment, below). This treatment is especially critical for shipping of winter-grown plants.

Growers should be aware that the overall quality and postharvest characteristics of winter-grown roses are generally inferior to those of crops grown in spring or early summer. However, there is a high consumer demand for potted roses during the winter, especially for Valentine's Day, and the potential for growth in sales at this time of the year is high. Paying particularly close attention to postharvest shipping and handling procedures for winter-grown crops will no doubt prevent a loss in consumer confidence that could damage future sales.

Flowers developed faster during shipping when plants were grown in the summer as opposed to the winter (Chen, 1990). Flowers on summer-grown plants developed at least one stage (see Stage of Development, below) during shipping, whereas flowers on winter-grown plants developed very little. This difference should be accounted for when shipping at different times of the year so that plants arrive at the proper stage of development for the marketing channel being used.

Temperature

L. Hendriks (personal communication, 1988) forced potted roses on a long-cycle schedule in the low-light conditions of northern Germany. Many combinations of production day/night temperature were used and shelf life was evaluated. The end of shelf life was defined to be when the rose leaves dried out and flowers abscised. The greatest shelf life was obtained when plants were grown with a 72°F/64°F (22°C/18°C) day/night temperature, while a 64°F/79°F (18°C/26°C) day/night temperature reduced shelf life about 60%. Pot roses should not be grown, therefore, in a higher night than day temperature.

Light

Although light is known to be very important for pot rose growth and development (see Chapter 5), little work has been done to investigate the effects of production lighting on postharvest longevity except in the area of light quality. Orange Sunblaze plants grown under a filter that greatly reduced the far-red light component from natural light had a slightly higher percentage of yellow leaves after simulated shipping than plants grown under water-filtered light, though overall postharvest quality was unaffected (Rajapakse and Kelly, 1993). Also, the leaf yellowing under the far-red light absorbing filter was associated with a reduction in leaf sucrose and starch concentration by 40% and 65%, respectively.

Sanitation

Plants must be free from insects and diseases. Plants with pests or diseases are of inferior quality and may infect other plants in a shipment. Damaged or infected plants also produce more ethylene, which further reduces quality. With low temperatures during storage and transport, potted plants are particularly susceptible to gray mold (*Botrytis cinerea*).

Cultivar Selection

The influence of genetics on the postharvest characteristics of potted roses is obvious in studies conducted to determine the shelf life of potted roses. Recent breeding programs with potted roses have not only focused on improved colors and pot performance, but also on evaluation of postharvest quality. Breeding programs should include an evaluation of every selection in an interior environment before it is released for production and marketing.

Halevy and Kofranek (1976) reported that Garnette and Orange Margot Koster plants had little bud and leaf abscission even at warm shipping temperatures, whereas Pink Margot Koster was severely injured. Red Sunblaze was found to last longer in an interior environment than Orange Sunblaze (Chen, 1990). Orange Sunblaze and Confection had more leaf abscission and/or yellowing than Red Sunblaze, Lady Sunblaze, Shy Girl, and Red Minimo after 4 days of simulated shipping at 46°F (8°C) and 7 days in a simulated interior environment (Clark et al., 1991b). Monteiro et al. (1992) found interior longevity to be longest for Red Sunblaze (23 days), followed in order of descending longevity by Orange Sunblaze (18 days), Lady Sunblaze and Candy Sunblaze (16 days), and Royal Sunblaze (13 days). Also, Dreaming Parade and Victory Parade exhibited greater longevity than Femini Rosamini and Red Minimo (A. S. Andersen, M. Serek, and P. Johansen, personal communication, 1994).

HARVEST CONSIDERATIONS

Unfortunately, there are no enforced standards for quality of potted floricultural crops in the U.S. or Europe, although many sets of recommendations exist. The most common method of determining plant standards is by pot size. Potted plants are also evaluated by the size of the plant, number of flowers or buds, leaf and flower color, and freedom from defects. Difficulties in establishing quality standards stem from the wide diversity of species and cultivars being produced and the uniqueness of each end-market such as mass merchandiser, discount store, florist shop, etc. General quality guidelines for potted roses are similar to those for other potted flowering crops, including a balance of plant size with pot size, sufficient flowering shoots to give an even and balanced floral display, flowering shoots of a uniform height, uniform flower opening, good development of flower color and form, and a healthy vigorous appearance of the flowers and foliage.

Stage of Development

Chen (1990) delineated the stages of development for potted rose flowers (Plate 20 and Table 11). There are usually several flower stages on the rose plant at the time of harvest which allows for a much longer floral display after harvest. Generally, potted roses should be sold when most of the flowers are in the bud stage. However, the degree to which the flowers should be open for marketing depends upon the marketing channel. Plants with one flower at stage 3 or above and with the majority of the buds showing color (stage 2) are most desired by mass market and garden center merchandisers. Traditional florist outlets want plants with maximum stage 3 flowers unless delivered close to a major holiday. For holidays, plants with an open flower (stage 4) are acceptable. If plants mature too early for a targeted market date, they can be held for a week or more by reducing greenhouse temperature to 45–50°F (7–10°C) when the buds are ready to open (stage 1).

Season of production and shipping temperature influence the degree to which flowers open during shipping. Flowers on summer-grown plants generally advance one stage during shipping while flowers on winter-grown plants developed only one third as much when shipped at 61°F (16°C) for 4 days (Chen, 1990). However, shipping plants at 40°F (4.5°C) slowed flower development significantly and is recommended whenever possible.

Chemical Treatments

Foliar chemical treatments applied just prior to shipping have been shown to be beneficial in reducing some postharvest disorders of potted roses. Application of 50 or 100 ppm of the cytokinin PBA [6(benzylamino)-9-(2-tetrahydropyranyl)-9H-purine] reduced flower bud and leaf abscission of the grafted potted rose varieties Garnette, Orange Margo Koster, and Pink Margot Koster (Halevy and Kofranek, 1976). Benzyladenine was found to have little effect on flower longevity when applied alone or in combination with the anti-ethylene agent silver thiosulfate (STS) on Orange Sunblaze and Red Sunblaze plants (Cushman et al., 1994). In contrast, Serek and Andersen (1993) reported that 180 ppm BA increased postharvest plant longevity of Victory Parade. Leaf chlorosis and abscission on Orange Sunblaze plants, which occurred after simulated shipping at 61°F (16°C) for 5 days, could be greatly reduced by a pre-shipping treatment of 25 ppm spray with the cytokinin benzyladenine (BA). The cytokinin treatment did not, however, reduce the number of etiolated shoots (Clark et al., 1991b). BA (100 ppm) was also found to reduce leaf yellowing on Belle Sunblaze but had little effect on flower-bud yellowing and opening (Tjosvold et al., 1994). If leaf yellowing is a problem, a 25 ppm BA or 50 ppm PBA treatment prior to shipping is recommended.

Exposure of Belle Sunblaze plants to 1 ppm ethylene has been shown to accelerate flower bud and leaf yellowing and abscission under interior conditions (Tjosvold et al., 1994). Serek (1993) also demonstrated the sensitivity of pot roses to ethylene by showing that 500 ppm ethephon, an ethylene-releasing compound, reduced floral and plant longevity and increased floral abscission on Victory Parade pot roses. Cushman et al. (1994) found that STS increased post-storage floral longevity on Orange Sunblaze and Red Sunblaze plants under simulated interior conditions after simulated shipping at 61°F (16°C) for 4 days. An application of 2 mM STS was found to be as good as 3 mM, but 4 mM resulted in phytotoxicity (Cushman et al., 1994). Flower bud and leaf abscission on Belle Sunblaze plants was prevented by 1 mM STS, but leaf yellowing was only partly reduced (Tjosvold et al., 1994). Application of 0.4 mM STS to Victory Parade plants increased flower longevity as much as sprays of up to 1.6 mM (Serek, 1993).

STS has a greater effect on flowers than leaves, and BA has the reverse effect. A combination of BA plus STS is, therefore, a good treatment for postharvest longevity, as BA and STS improve leaf and flower retention, respectively. The concentrations needed, however, will depend upon the cultivar being treated. Also, BA will improve flower longevity of some cultivars, such as Victory

Parade, but not others, such as Orange Sunblaze and Red Sunblaze.

STS can be prepared from commercially available formulations. As with any chemical application, label directions should be carefully followed and a small number of plants treated to determine phytotoxicity prior to treating an entire crop. Solutions should always be made fresh and different compounds applied separately, allowing the foliage to dry between applications. Foliage must be dry prior to sleeving and boxing to prevent leaf burn. Excess solution must be disposed of properly according to label directions.

Care Tags

Each plant should have a care tag attached that gives cultural information for consumers. The tag should include information such as the cultivar name and the requirements for light, fertility, watering, and temperature. The more information the grower can provide consumers, the greater the chances that the consumer will succeed with the product. Consumer education is an important aspect of improving consumer satisfaction.

Packaging

The type of packaging materials used for shipping depends on many variables such as end market, distance to market, size of plants, and grower preference. Potted roses are often sleeved before they are shipped. Plant sleeves protect leaves, flower buds, and stems from physical damage during shipping and handling. Sleeves are made of paper, mesh, plastic, or fiber materials. The type of sleeving material does not matter, but extending the sleeve 2 inches (5 cm) above the rose flower buds is recommended for adequate protection.

Smaller sizes of potted roses are often boxed when they are shipped. It is critical that the plants fit properly in the box and that the box can withstand the rough handling often encountered in shipping. Recommended standards for box sizes have been made by the Society of American Florists and the Produce Marketing Association (Table 12). Plants should be held in a cool place after they are boxed because heat buildup within the box will reduce quality (Maxie et al., 1974). The box should be clearly labeled with the temperature limits for shipping and "This End Up" and "Live Plants" designations.

Storage and Shipping

While most U.S. companies box finished roses for shipping, European producers avoid many problems associated with postharvest handling of potted roses by shipping them for shorter times and using open carts that allow for good air circulation around the plants during shipping and storage (Plate 21). If potted roses are to be sleeved and boxed, they should be transported to a cool location for processing.

Ethylene gas is one of the major causes of premature flower senescence, leaf yellowing, abscission, and reduced plant quality. To minimize ethylene damage, good ventilation in storage areas is essential. One air exchange per hour is usually sufficient if low levels of ethylene are occasionally present. Cool (40°F, 4.5°C) holding temperatures must be maintained because plants held at lower temperatures are less responsive to ethylene. Treating plants with STS prior to shipping will minimize the effects of ethylene. Also, potted roses should not be stored or shipped with fruits, vegetables, or cut flowers and should be kept away from combustion engines because of the danger of ethylene.

The combination of dark and warm temperature is known to promote abscission of flowers and leaves in roses. Exposure of pot roses to these conditions during shipping can result in excessive etiolated shoot growth in some cultivars (e.g. Red Minimo), leaf yellowing and abscission, and abnormal flower opening (Plate 18) and color development (e.g. blueing of Orange Sunblaze flowers and a darkening of Red Sunblaze flowers) (Cushman et al., 1994; Clark et al., 1991b). When marketable Mother's Day potted roses were held in boxes in sunlight for only 4 hours, outside air temperature increased from 72 to 85°F (22 to 29°C), media temperature inside the box rose from 60 to 81°F (15.5 to 27°C) and air temperature inside the box rose from 73°F (23°C) to as high as 104°F (40°C) (Maxie et al., 1974). When plants were removed from the box and evaluated, no damage was noted on open flowers, but unopened buds abscised before they opened and plants were judged unacceptable for sale.

Shipping temperature and duration are also critical. Halevy and Kofranek (1976) demonstrated that storage at 68–72°F

(20–22°C) for 6 days resulted in little or no bud and leaf abscission on Garnette and Orange Margo Koster plants when removed from the boxes, but Pink Margo Koster plants were severely affected. However, plants of all 3 cultivars had severe leaf senescence by 10 days after shipping. Storage at 34–37°F (1–3°C) for 6 days prevented this flower bud and leaf abscission. Shipping at 34–37°F (1–3°C) for 1 day followed by 1 day at 68–72°F (20–22°C) and 3 more days at 34–37°F (1–3°C) did not adversely affect plants. Heins (1981) noted that storage of Garnette potted roses for 10 days at 35°F (1.5°C) did not adversely affect postharvest life of the plants. When plants were stored longer, however, flowers developed a "water-soaked" appearance and did not open properly. Chen (1990) found that storage of Orange Sunblaze and Red Sunblaze plants at 39°F (4°C) for up to 6 days or at 61°F (16°C) for up to 4 days could be used for maintaining plant quality after shipping. However, shipping at 82°F (28°C) should be avoided altogether (Plate 19). Also, Red Sunblaze was more tolerant to shipping than Orange Sunblaze (Chen, 1990; Clark et al., 1991b). Nell and Noordegraaf (1991) found that, after 3 weeks of interior evaluation, the number of open flowers on Orange Rosamini plants decreased as shipping temperature increased from 41 to 63°F (5 to 17°C); no effect was correlated with a shipping duration of up to 9 days. Plants held for 9 days at 63°F (17°C) had the fewest buds showing color per plant. In the same study, Favorite Rosamini, Sweet Rosamini, and Golden Rosamini were found to be more sensitive to shipping than Orange Rosamini, Red Rosamini, and White Rosamini. To summarize all of these studies, the longer the duration of shipping, the lower the temperature required for good results. Shipping for no longer than 4 days while maintaining a temperature as close to 40°F (4.5°C) as possible is recommended.

RETAIL AND CUSTOMER HANDLING

Potted roses should be inspected when they arrive. Samples should be removed from each shipment, and foliage as well as the root system should be carefully inspected for defects. If plants are damaged, contact the supplier immediately or file a claim with the shipping company. Under no circumstances should a shipment of flowering potted rose plants be allowed to freeze. When possible, plants should be removed from boxes, unsleeved and placed in high light but not in direct sunlight. The longer a plant stays in a dark box the greater the reduction in quality. Plants should also be kept out of drafts from heating and air conditioning vents and sources of ethylene.

Light

As mentioned above, light is the most important factor for rose-plant growth and development. This holds true for handling in retail and consumer environments. Nell and Noordegraaf (1992) found that after shipping for 3 days at 41°F (5°C), there were more open flowers on Orange Rosamini plants held for 3 weeks in a simulated consumer environment under 4 W per square meter (126 footcandles or 18.5 µmol per second per square meter) of cool white fluorescent light than on plants held under 1 W per square meter (32 footcandles or 4.5 µmol per second per square meter). When plants were exposed to a simulated retail environment prior to being placed under the simulated consumer environment, the number of open flowers was reduced when compared to plants that were not given the simulated retail treatment. Although information concerning minimum light levels is lacking for other cultivars, providing adequate light levels necessary to maintain plant quality throughout the retail handling and consumer use phase is critical to successful marketing of pot roses.

Temperature

Potted roses should be kept as cool as possible in the retail shop as well as in the consumer environment. A temperature of 45 to 60°F (7 to 15.5°C) is ideal but may not be practical. As the temperature increases, however, flower development accelerates and shelf life is decreased.

Watering

Potted roses are among the floricultural crops considered most susceptible to injury from drying of the potting media. Roses should be watered well immediately after unpacking, and watering should be carefully monitored throughout the display time in retail. If the medium dries, the salinity of the soil solution increases and foliar damage such as marginal chlorosis and necrosis and leaf abscission are likely to occur. Fertilizing potted roses in the retail shop is usually not necessary.

Garden Use

Potted roses lend themselves to transplanting to the garden after the indoor decorative life is finished in many parts of North America. Care tags with cultural information should be provided if this use is to be promoted. In the garden, potted roses need the same care as any other garden-type rose. This includes planting where the plants will receive at least a half day of full sunlight, adequate watering, regular fertilization during the growing season, regular spraying to prevent fungal diseases such as black spot and powdery mildew, regular removal of spent flowers, and severe pruning limited to early spring. Although transplanting potted roses into the landscape is possible, attempts to hold plants purchased for Valentine's Day until spring may be difficult in the home unless very high light intensity and adequate media moisture and air humidity can be supplied. Heins (1981) has suggested that care tags for potted roses to be sold during the winter carry the following statement on the care tag: "This rose has been grown for your midwinter enjoyment. One should enjoy it now. Attempts to hold it indoors until spring may not be successful."

SUMMARY

Although potted roses have great market potential, postharvest handling of the crop must improve if it is to reach its full potential. The following recommendations should serve as guidelines for helping to insure consumer satisfaction:

- Keep plants well irrigated during production and in all phases of postharvest handling, including retail and consumer display areas.
- If using a constant liquid feed system, 200–300 ppm N from a balanced fertilizer is recommended with a clear water leaching every 2 to 3 waterings. Prior to sleeving and shipping, media should be leached with clear water.
- If CO_2 enrichment is used during production, plants must be shipped at low temperatures and for short durations to prevent etiolated shoot growth.
- Avoid producing roses during mid-winter without adequate light or during the hottest summer periods without adequate

cooling of growing facilities. Moderate day/night temperatures (72/64°F or 22/18°C) during forcing are recommended for longest postharvest performance.

- Spring and early summer production is usually optimal for postharvest longevity. However, this recommendation is highly dependent on location.
- Keep crops free from pest problems.
- Select cultivars carefully according to growing region and postharvest characteristics.
- Harvest pot roses when most of the flowers are in stage 2 (Plate 20 and Table 11).
- Supply a care tag with each potted rose.
- Sleeve roses before shipping and box if necessary to prevent physical damage during shipping.
- Roses should be shipped and stored for as short a time as possible. Ideal storage temperature is 35–40°F (1.5–4.5°C).
- Avoid exposing plants to any source of ethylene gas, including combustion engines, fruits, vegetables, and cut flowers.
- BA (25 ppm) or PBA (50 ppm) applied as a pre-shipping spray reduces leaf chlorosis if warm temperatures or long storage are anticipated and will increase flower longevity for some cultivars.
- STS (adjust concentration depending on cultivar) applied as a pre-shipping spray will substantially reduce flower abscission that occurs during shipping and interior use.
- Providing minimum light levels needed for maintenance of plant quality is critical for successful marketing and consumer use.

8 Production Budget

Cost of production is a critical component in making decisions concerning greenhouse product mix. Boyd et al. (1982) developed production budgets for several greenhouse crops. Brumfield et al. (1981) produced an excellent study utilizing industry survey data to determine actual fixed costs per square foot for large, medium, and small wholesale greenhouse firms. They defined costs based on whether the crop was to be distributed through retail florists or mass market outlets. Vaut et al. (1973) and Boyd and Phillips (1983) developed guidelines for linear programming of greenhouse crops, which helps determine optimal product mix for growers.

The cost of producing any greenhouse crop varies among producers and across geographic regions. It is important, therefore, for each producer to determine his/her own production costs so that crop costs may be determined accurately and a reasonable profit may be budgeted. The Professional Plant Growers' Association surveys growers and publishes a report on greenhouse operating costs (Anonymous, 1987).

The potted rose budget developed by Kelly (1990) and shown in Tables 13–17 was based partially on overhead cost data collected by Brumfield et al. (1981) for medium-size greenhouse operations. These data were adjusted for inflation, and a computer-budgeting program for greenhouse crops developed by Kelly et al. (1990) was used to calculate fixed and variable costs for potted rose production.

A number of assumptions were made in developing the budget. The crop was assumed to be 2000 patented roses grown short-cycle in 4.5-inch (11-cm) pots on their own roots from purchased liners (Table 17A). A late March planting date was chosen and a 5% crop loss was assumed. Plants were potted and initially spaced pot-to-pot on the bench. After 2 weeks, plants were pinched and spaced at 8 inches on center. Standard fertilization and pesticide application were budgeted as described in Chapters 5 and 6. Plants were assumed to be marketed with a 30% markup after 6 weeks at the second spacing.

Perhaps the most difficult component of production costs to calculate is overhead costs. There is no standard that fits all growers. Overhead costs vary by geographic area, management strategy, and size and type of operation. Fixed overhead costs include land and depreciable items such as buildings and equipment. Varying overhead costs are management salaries, utilities, taxes, and miscellaneous expenses.

For this study, a gutter-connected, 20,000-square-foot, double-layered polyethylene greenhouse with 15,400 square feet of bench space was assumed as the size of the total operation. Other assumptions were budgeted as indicated in Table 13. Standard depreciation schedules are presented in Table 14. Table 15 was developed from the data of Tables 13 and 14. The annual costs per square foot of greenhouse and per square foot of bench space were estimated to be $0.89 and $1.15, respectively. Next, management salary, insurance, utilities, etc. were established for the 20,000-square-foot facility (Table 16). Total annual overhead and building and equipment costs per square foot of greenhouse and per square foot of bench space were estimated to be $4.11 and $5.34, respectively.

Only 267 square feet of bench space were occupied by the potted roses for the 2-week phase of pot-to-pot spacing, but 844 square feet were needed for the final 6 weeks of production, thus requiring 2% and 5% of the total greenhouse square footage, respectively (Table 17B). Total fixed costs for the potted rose crop was estimated to be $0.24 per pot.

Variable costs account for all the inputs specific to the crop, including shipping costs for the required production materials (Table 17C). Generally, material costs are relatively easy to calculate because they are the costs directly related to the amount of materials and supplies purchased for the crop being produced. Material costs vary directly by the number of pots being produced. Another component of variable costs, labor costs, is rather difficult to assign to a crop, and it varies substantially depending on the amount of labor-saving devices that a grower has installed in his/her greenhouse operation. Labor involved in the production of potted roses includes the time spent unpacking, repotting, and spacing the plants on the bench; pinching and respacing; spraying, fertilizing, irrigating, and harvesting; and packaging and shipping.

When variable costs were considered for potted roses, the 2000-pot crop had a variable cost of approximately $2,498 for materials and labor, or $1.25 per pot. Thus, the total cost of production (including overhead) was $2,961 or $1.49 per pot (Table 17D). With a 30% markup the potted roses should wholesale for $1.94 per pot.

Overhead costs (Table 16) for 4.5-inch (11-cm) potted roses were estimated to be about 16% of total potted rose production costs. This is much lower than the overhead for many floricultural crops, including bedding plants, which are assumed to have low overhead costs of 25% of total costs (Faber et al. 1986). The low percentage of overhead costs for potted roses primarily results from the short time that a short-cycle rose is held on the bench and to the relatively high variable costs for the crop. Variable costs account for approximately 84% of the production cost of a potted rose; 62% of the variable cost is from the cost of purchased liners.

Supplemental high-intensity lighting is increasingly used to improve the quality of plants produced from late fall to early spring, particularly for short-cycle-grown crops. The additional variable cost associated with lighting from high-intensity discharge fixtures is shown in Table 18. These figures do not include the fixed cost for the fixtures themselves, but give cost estimates for several different electric use costs. Use costs will vary greatly from location to location and can be influenced by the availability of quantity discounts, off-peak hour rates, or other rate-reducing formulas. Producers should be familiar with local electrical rate structures to realize savings. The estimates in Table 18 are based upon lighting for 16 hours per day with 500 foot candles of light. Producers located at a latitude with winter sunshine above 500 foot candles may only need to light for 8 hours at the end of each sunny day to get the same results. Other ways of reducing the need for supplemental lighting are discussed in Chapter 5.

9 Marketing

Marketing is the key to successful floriculture crop production. Regardless of the number or quality of pot roses grown, the crop is not a success until it is sold. On a tour of a large greenhouse operation, a comment was made to the owner about the beauty of the 1+ acres of a finished flowering crop. The owner then replied that, yes, the crop was beautiful, but even more beautiful was the site of the empty greenhouse after the crop had been sold!

Selling a pot rose crop requires just as much effort as growing it. Successful marketing depends on several factors, including the time of year, type of product, color mix, and market channel.

TIME OF YEAR

Traditionally, pot roses have been a Mother's Day crop, although they were also marketed for Easter (MacKay, 1985). Production of pot roses consisted almost entirely of greenhouse forcing of field-grown plants of the Garnette and Koster cultivars. Growers tried to develop a market for Valentine's Day with the traditional field-grown cultivars (Asaoka and Heins, 1982; Heins, 1981), but because of the need for early digging of field-grown plants, the cost of the larger pot sizes required for bare-root plants, and the cost of supplemental lighting, winter forcing remained limited. Based on work of Moe (1973), growers in Denmark and Holland developed an economical system with cuttings and supplemental light for year-round production in small pots for mass market sales. With the recent development of new, better-adapted pot cultivars and improved greenhouse technology, pot roses can be economically produced year-round to satisfy an increasing demand.

Although pot roses are presently grown year-round by a few, highly specialized growers, the majority of North American producers still grow the bulk (80%) of their pot rose crop for the spring holidays from Valentine's Day to Mother's Day. Recent trends show an extension of the spring market later into the season. Many growers are reporting good sales in June, while some have found an expanding fall market from early September until Thanksgiving. Two periods during which sales are not well established are summer (July–August) and late fall (Thanksgiving to Christmas). Even the large, specialized pot rose producers who grow the crop year-round sell most of the crop between January and October and produce little for Thanksgiving and Christmas.

TYPE OF PRODUCT

Pot roses are not simply a single item such as a 4-inch (10-cm) pot, but rather a complete line of different products from which each grower can choose the type best suited for a particular production system and market strategy (Plates 3, 22, and back cover, *top*). The greatest number of plants grown are in 4-inch 10-cm) pots (cover photo). However, because of highly competitive pricing, 4-inch pots are mostly grown by large, specialized growers who can produce sufficient volume to realize the economies of scale needed to justify the investment required. The most competitive producers in this category are taking advantage of sophisticated greenhouse technologies such as moveable benches, high-intensity lighting, and automated potting and plant-moving equipment, all of which require a large capital investment. Although growers who buy prefinished plants in 4-inch (10-cm) pots can also finish them in the same container, the 4.5-inch (11-cm) pot is preferred by growers who are finishing liners. It provides a larger product and a higher return than a 4-inch pot without tying up significantly more bench space. The 4.5-inch (11-cm) pot is the smallest size recommended for the large-flower/large-leaf miniature cultivars such as the Sunblaze series.

Six-inch pots (15-cm) are most widely grown for specific holidays, but some growers specialize in year-round production of 6-inch (15-cm) pots to complement their 4-inch (10-cm) pot program. Six-inch (15-cm) pots are also commonly used when repotting prefinished 4-inch (10-cm) potted plants. Seven and 8-inch

(17.5 and 20-cm) diameter pots are used for forcing bare-root plants and patio trees. They represent the "top of the line," and are sold mostly for Mother's Day.

In Europe, 2 other products are widely available. The 2.5-inch (6-cm) "personal mini rose," called "Kuvertrose" in Denmark, is very popular, especially in Germany and northern Europe. At the other end of the spectrum are the 12 and 15-inch (30 and 37.5-cm) window boxes, which are planted with 2 and 3 plants, respectively, and sold throughout the spring in France and Italy (back cover, *top*). These are either produced by the greenhouse producer or assembled with surplus plant material by the garden center retailer. Other products include hanging baskets and trellis-trained plants.

Another potential use of pot roses is for planting into the garden or patio containers after flowering is finished (Plate 22). Many consumers want to do this, especially when the plants are purchased at a frost-free time of year. In many climates these roses perform well, and this additional use can be promoted with great success. However, in areas with hot, humid summers such as the southern U.S., many pot-forcing cultivars do not perform as well as those selected primarily for garden use. Also, many pot rose cultivars are not winter hardy in the northern U.S. and Canada. In such a situation, garden use should be promoted cautiously, if at all. Local trials to determine the best pot cultivars for garden use are recommended. In addition, cultural notes including disease control instructions should be included for all garden use promotions. Colored care tags are an effective way to convey this information.

COLORS

Thanks to continuing breeding efforts, a complete palette of colors in all types of pot roses are now available. However, the typical color mix that a grower should consider producing is still 50% red (dark red and vermillion), 30% pink, 15% yellow, and 5% white. Such a mix is not significantly different from the typical cut-flower rose palette grown in North America. The pastel and other soft colors are, however, taking a larger part of the market in northern Europe, where consumers demand new colors and shades. In general, stronger colors seem to be preferred in North America, whereas pastels are popular in Europe, especially in northern areas.

In North America, red-flowered cultivars account for 60 to 80% of Valentine's Day production. For Easter and Mother's Day, a wider range of colors is preferred, although red and orange-red still predominate. In late spring and early fall, strong colors that will not fade, such as orange and hot pink, are preferred.

When informally surveyed about flower color, growers answer that a bicolor and a lavender are needed. Most growers are not looking for new cultivars with better or different colors, however, but for better plant performance. Sales representatives, on the other hand, feel that the availability of a full range of colors is critical to successful sales. Unfortunately, little has been documented about consumer color preferences in North America.

MARKET CHANNELS

Historically, shipping problems have limited pot rose sales to relatively local markets. With the development of cultivars with better postharvest life and the use of shipping agents such as silver thiosulfate, growers are now able to ship across large distances. Because of these recent developments, the regional and national distribution channels for pot roses in North America are now similar to those of the major flowering pot plants such as poinsettias and chrysanthemums. However, pot roses are still considered by the trade as one of the "other flowering plants" such as kalanchoes and exacum. There are few specialized growers of this crop in North America and none to our knowledge who are growing them as their only crop. However, a recent trend in America is the emergence of large firms that specialize in year-round pot production similar to the "plant factories" in northern Europe. In Europe, pot rose production figures are considered in a separate major crop category.

In Europe, the market channels vary according to each country. Exporting countries such as Denmark and Holland rely on their marketing cooperatives (Dutch Auctions or Veilings and Danish GASA) to sell the product. Growers in these countries are not usually directly involved in selling. They can focus on the growing end of the business and are continuously developing new technologies often beyond the means of the average grower in the other countries. In fact, the production of several million units in northern Europe is concentrated in a handful of growers who more or less control the supply to the co-ops. Because of the scale of production, competitive pricing, and highly sophisticated distribution systems, the GASA and Veilings are able to supply pot roses to European chain stores, mass marketing outlets, and food stores usually more cheaply and sometimes faster than local growers.

In European countries other than Holland and Denmark, growers are also marketers. Each has their own set of customers, and they usually compete among themselves as well as against the product from the GASA and Veilings. Because they normally cannot compete in price and scale, the successful growers are usually niche marketers who position themselves in a segment of the market not covered by the exporting countries. This means growing a product not offered by the large growers. These include larger pots, specialty items such as finished window boxes, and different colors and cultivars. They can compete in this way because the main drawback of the sophisticated export-driven Dutch and Danish systems is its lack of flexibility. In fact, it is so rigid that only a handful of products and cultivars can be produced. For the grower facing this kind of competition, there is an opportunity to offer a different product that can be more adapted to the local taste than the basic mass-produced 4-inch (10-cm) potted rose.

Such niche growers usually sell their product to the higher end of the market such as the upscale chain stores and garden centers as well as the retail florists where point of purchase material can be used to increase sales. For example, a very successful European grower is using an interesting strategy with chain-store customers. An attractive display of the product, mostly 6 and 8-inch (15 and 20-cm) pots complete with cultural tips for the consumer, are framed with a few patio rose trees in front to attract the eye of the consumer. The patio trees provide the impulse appeal that, in turn, helps to sell the 6-inch (15-cm) product.

Most producers in North America are also marketers. Though not based on cooperatives, large growers tend to sell to the mass market, while smaller growers service more traditional outlets. The relationships between production strategy and marketing channel seen in the well-developed European pot rose market provide valuable insight for American growers, both large and small, who are entering the rapidly expanding North American market.

In conclusion, the grower who is planning a pot rose crop should first ask where and when the crop is going to be marketed. As with any crop, the marketing strategy must be defined before sticking the first cutting or buying the first liner. This is the only approach for successful crop production in floriculture today.

10 Future Considerations

The future for pot rose production appears bright. Production in the United States and Canada has more than doubled between 1989 and 1995 and is expected to continue to rise. Nonetheless, as with any floricultural crop, there is room for improvement.

In most of North America, high-intensity supplemental lighting is needed for quality pot rose production, especially during the winter. Unfortunately, this is a practice that carries a high price tag. Improved efficiencies such as optimizing the lighting during the early stages of establishment and forcing, followed by finishing under natural light, would help reduce the impact of lighting on production costs.

The increasing concern for responsible environmental stewardship and the promise of cost efficiencies press for the development of nonchemical height-control methods. The use of spectral filters for altering light quality as a means of controlling plant growth will grow in importance as technological advances are made in production of greenhouse covers. This area has a bright future, and our knowledge is continuing to grow.

Another element of pot rose culture that will no doubt be improved is postharvest performance. Our knowledge of how to improve our product for consumers has increased greatly over the past few years, yet we have only begun to discover how the production environment and post-production handling practices can be used to ensure postharvest longevity.

Even with improvements in production and handling, however, the main contribution that will drive the expansion of pot rose production will continue to be the introduction of new cultivars. Improvements will be made in 3 major areas: (1) greenhouse performance, (2) postharvest performance, and (3) garden performance. Although currently there is no pot rose cultivar that is considered superior in all 3 of these areas, the breeding of pot roses specifically for greenhouse culture and consumer enjoyment in the home and garden is in its infancy. The acceleration in the introduction of new pot rose cultivars continues unabated, so the next "miracle" cultivar is probably just around the corner.

The cornerstone of improving production efficiency and plant improvement through breeding is our understanding of the fundamental life processes that underlie rose plant growth. Without continued research into the physiology and genetics of rose plant growth, forward momentum for improving pot rose production will cease. As pot rose production continues to grow, the justification for this research grows as well. The establishment of a broad research base will sustain the industry well into the future.

11 Tables

Table 1. Rose cultivars used for pot forcing in North America (adapted from Hammer, 1992). Except as noted, the registered name is the name officially registered with the American Rose Society as published in Modern Roses 10 (Cairns, 1993) and the Combined Rose List 1994 (Dobson and Schneider, 1994) or is the name found in commerce in North America. The American Rose Society is the International Registration Authority for Roses. The company controlling U.S. propagation rights of a pot rose cultivar either owns the patent, owns the trademark of the common name, has been granted exclusive propagation rights in a territory that includes the U.S., or a combination of the three.

REGISTERED NAME (COMMON NAME)	COMMERCIAL SYNONYMS	DENOMINATION NAME	U.S. PLANT PATENT NUMBER	DATE OF ORIGINATION OR INTRODUCTION	FLOWER COLOR	COMPANY CONTROLLING U.S. PROPAGATION RIGHTS
POLYANTHA						
Dick Koster				1929	Deep pink	
Margo Koster	Sunbeam			1931	Salmon	
Mothersday	Fêtes des Mères			1949	Deep red	
	Morsdag					
	Muttertag					
Orange Koster				Unknown-NR[a]	Orange	
Orange Margot Koster				Unknown-NR	Orange	
Pink Margot Koster				Unknown-NR	Pink	
Tammy			3464	1972	Light pink	

REGISTERED NAME (COMMON NAME)	COMMERCIAL SYNONYMS	DENOMINATION NAME	U.S. PLANT PATENT NUMBER	DATE OF ORIGINATION OR INTRODUCTION	FLOWER COLOR	COMPANY CONTROLLING U.S. PROPAGATION RIGHTS
Triomphe Orléanais				1912	Cherry red	
White Koster	Blanche Neige White Dick Koster			1929-NR	White	
FLORIBUNDA						
Bright Pink Garnette				Unknown-NR	Pink	
Carol Amling	Carol		1126	1953	Deep rose-pink, edged lighter	
Garnette	Red Garnette			1951	Garnet-red, base light lemon yellow	
Golden Garnette			1898	1960	Golden yellow, edged lighter	
Marimba				1965	Medium pink	
Melodee				Unknown-NR	Yellow	
Minuette	Laminuette		3162	1969	White, edged red	
Roswytha				1968	Pink	
Thunderbird			1677	1958	Rose-red	
MINIATURE—SMALL FLOWERED						
Chipper[b]			2764	1966	Salmon-pink	The Conard-Pyle Co.[c]

Table 1 continued.

REGISTERED NAME (COMMON NAME)	COMMERCIAL SYNONYMS	DENOMINATION NAME	U.S. PLANT PATENT NUMBER	DATE OF ORIGINATION OR INTRODUCTION	FLOWER COLOR	COMPANY CONTROLLING U.S. PROPAGATION RIGHTS
Cinderella			1051	1953	Satiny white, tinged pale flesh	
Confection		JACute	6518	1988	Medium pink	Bear Creek Gardens
Day Glow[b]		JACrink	6795	1989	Deep pink	Bear Creek Gardens
Minimo[b] Series						
Dainty Minimo[b]		RUIfifty	6237	1986-NR	Pale pink	Bear Creek Gardens
Karmina Minimo[b]		RUInielov	7540	1991-NR	Crimson red	Bear Creek Gardens
Pink Minimo[b]		RUIpiko	5768	1984-NR	Medium pink	Bear Creek Gardens
Red Minimo[b]		RUImired	5770	1987	Dark red	Bear Creek Gardens
Rosy Minimo[b]		RUInanny	PPAF[d]	1989-NR	Pink	Bear Creek Gardens
Mini-Wonder[b] Series						
Crimson Mini-Wonder[b]		MEInochot	8242	1991-NR	Dark red	The Conard-Pyle Co.
Pink Mini-Wonder[b]	Pink Minijet[b]	MEIselgra	PPAF	1990	Medium pink	The Conard-Pyle Co.
Orange Mini-Wonder[b]		LAVjack	7326	1990-NR	Orange	The Conard-Pyle Co.
Sweet Mini-Wonder[b]		MEIlepo	8504	1991-NR	Medium pink	The Conard-Pyle Co.
White Mini-Wonder[b]	Spot Minijet[b]	MEIzogrel MEIsogel	7276	1988	White	The Conard-Pyle Co.

REGISTERED NAME (COMMON NAME)	COMMERCIAL SYNONYMS	DENOMINATION NAME	U.S. PLANT PATENT NUMBER	DATE OF ORIGINATION OR INTRODUCTION	FLOWER COLOR	COMPANY CONTROLLING U.S. PROPAGATION RIGHTS
Yellow Mini-Wonder[b]	Spicy Minijet[b] Potluck[b] Yellow	LAVglo	4136	1985	Yellow	The Conard-Pyle Co.
Parade[b] Series						
Bianca Parade[b]		POULbian	PPAF	1994-NR	White	Bear Creek Gardens
Dreaming Parade[b]		POULoral	9018	1994-NR	Salmon pink	Bear Creek Gardens
Fame Parade[b]		POULtory	PPAF	1994-NR	Red	Bear Creek Gardens
Harmony Parade[b]		POULming	9017	1994-NR	Champagne pink	Bear Creek Gardens
Lavender Parade[b]		POULminet	PPAF	1991-NR	Lavender	Bear Creek Gardens
Pink Parade[b]		POULcar	7999	1991-NR	Pink	Bear Creek Gardens
Queen Parade[b]		POULann	8943	1994-NR	Pink	Bear Creek Gardens
Royal Parade[b]		POULspor	8942	1994-NR	Pink	Bear Creek Gardens
Starlight Parade[b]		POULstar	9016	1995-NR	White	Bear Creek Gardens
Sunset Parade[b]		POULgelb	PPAF	1994-NR	Yellow	Bear Creek Gardens
Victory Parade[b]		POULvic	8012	1991-NR	Medium red	Bear Creek Gardens
Peter's Dream				1989-NR	Medium red	
Pixie	Little Princess Princesita		408	1940	White, center light pink	
Potluck Gold[b]		LAVgold	8030	1991	Bright yellow	Bear Creek Gardens
Purple Passion[b]		JACpupot	PPAF	1994-NR	Mauve-pink	Bear Creek Gardens

Table 1 continued.

REGISTERED NAME (COMMON NAME)	COMMERCIAL SYNONYMS	DENOMINATION NAME	U.S. PLANT PATENT NUMBER	DATE OF ORIGINATION OR INTRODUCTION	FLOWER COLOR	COMPANY CONTROLLING U.S. PROPAGATION RIGHTS
Rainbow Series						
Rainbow Hot Pink [b]		DEViente	PPAF	1989	Deep pink	DeVor Nurseries, Inc.
Rainbow Surprise [b]		DEVpresa	PPAF	1989	Apricot blend	DeVor Nurseries, Inc.
Red Imp	Maid Marion Mon Tresor		1032	1951	Deep crimson	
Rise 'n' Shine	Golden Meillandina [b]		4231	1977	Rich medium yellow	Sequoia Nursery
Rosamini [b] Series						
Coral Rosamini [b]		RUIforto	6244	1988-NR	Coral	Bear Creek Gardens
Flaming Rosamini [b]		RUIflami	PPAF	1987-NR	Red-orange	Bear Creek Gardens
Golden Rosamini [b]		INTERgol	7430	1990-NR	Deep yellow	Bear Creek Gardens
Orange Rosamini [b]		RUIseto	6236	1988	Orange	Bear Creek Gardens
Pink Rosamini [b]		RUInidan	8338	1993-NR	Pink	Bear Creek Gardens
Red Rosamini [b]		RUIredro	5976	1987	Dark red	The Conard-Pyle Co.
Ruby Rosamini [b]		INTERsept	8349	1993-NR	Red	Bear Creek Gardens
Scarlet Rosamini [b]		RUIrupo	PPAF	1991-NR	Scarlet-orange	Bear Creek Gardens
Sweet Rosamini [b]		RUImarso	6243	1986-NR	Light pink	Bear Creek Gardens
Scarlet Gem	Scarlet Pimpernel	MEIdo	2155	1961	Orange-scarlet	
Shy Girl		JACwhim	6514	1988	White	Bear Creek Gardens
Starina [b]		MEIgabi MEIgali	2646	1965	Orange-scarlet	The Conard-Pyle Co. [c]

REGISTERED NAME (COMMON NAME)	COMMERCIAL SYNONYMS	DENOMINATION NAME	U.S. PLANT PATENT NUMBER	DATE OF ORIGINATION OR INTRODUCTION	FLOWER COLOR	COMPANY CONTROLLING U.S. PROPAGATION RIGHTS
Sweet Fairy			748	1946	Light pink	
MINIATURE—LARGE FLOWERED						
Amber Flash[b]		WILdak	5271	1982	Yellow-orange blend	The Conard-Pyle Co.
Apache Princess		TWOmin		1989	Orange-red	
Debut[b]		MEIbarke	6791	1988	Cream/Red bicolor	The Conard-Pyle Co.
Fantasy				1989-NR	Salmon pink	
Festival[b] Series						
Pink Festival[b]		LAVquest	8031		Blush pink with deep pink eye	Bear Creek Gardens
Purple Festival[b]		LAVpurr	PPAF	1992	Medium purple	Bear Creek Gardens
White Festival[b]	Springwood[b] White[e]	LAVsnow	PPAF	1991	White	Bear Creek Gardens
Yellow Festival[b]		LAVling	PPAF	1991-NR	Canary yellow	Bear Creek Gardens
Galaxy[b]		MORgal	4680	1980	Dark red	Sequoia Nursery
Hit[b] Series						
Absolute Hit[b]		POULrouge	PPAF	1995-NR	Bright orange	Bear Creek Gardens
Fantasy Hit[b]		POULdyb	PPAF	1995-NR	Dark red	Bear Creek Gardens
Nova Hit[b]		POULavon	PPAF	1995-NR	Soft pure pink	Bear Creek Gardens

Table 1 continued.

REGISTERED NAME (COMMON NAME)	COMMERCIAL SYNONYMS	DENOMINATION NAME	U.S. PLANT PATENT NUMBER	DATE OF ORIGINATION OR INTRODUCTION	FLOWER COLOR	COMPANY CONTROLLING U.S. PROPAGATION RIGHTS
Perfect Hit [b]		POULfect	PPAF	1995-NR	Salmon	Bear Creek Gardens
Pure Hit [b]		POULdel	PPAF	1995-NR	White	Bear Creek Gardens
Sunblaze [b] Series						
Candy Sunblaze [b]	Romantique Meillandina [b] Concertino Romantic	MEIdanclar	7621	1991	Deep pink	The Conard-Pyle Co.
Cherry Sunblaze [b]		MEIbokarb MEIbekarb	PPAF	1993	Bright red	The Conard-Pyle Co.
Classic Sunblaze [b]	Duc Meillandina [b] Duke Meillandina [b]	MEIpinjid	PPAF	1985	Hot pink	The Conard-Pyle Co.
Golden Sunblaze [b]		MEIlipo MEIcupag	8493 PPAF	1992-NR	Medium yellow	The Conard-Pyle Co.
Gypsy Sunblaze [b]		MEImagul	PPAF	1994	Red/Yellow bicolor	The Conard-Pyle Co.
Lady Sunblaze [b]	Lady Meillandina [b] Peace Meillandina [b] Peace Sunblaze [b]	MEIlarco	6170	1989	Light pink	The Conard-Pyle Co.
Magic Sunblaze [b]		SCHanbiran	5101	1984-NR	Red blend	The Conard-Pyle Co.
Orange Sunblaze [b]	Orange Meillandina [b]	MEIjikatar	4682	1981	Orange-red	The Conard-Pyle Co.
Pink Sunblaze [b]	Pink Meillandina [b,e]	MEIjidiro	4961	1980	Pink	The Conard-Pyle Co.

REGISTERED NAME (COMMON NAME)	COMMERCIAL SYNONYMS	DENOMINATION NAME	U.S. PLANT PATENT NUMBER	DATE OF ORIGINATION OR INTRODUCTION	FLOWER COLOR	COMPANY CONTROLLING U.S. PROPAGATION RIGHTS
Red Sunblaze[b]	Prince Meillandina[b,e] Prince Sunblaze[b]	MEIrutral	7021	1988	Dark red	The Conard-Pyle Co.
Royal Sunblaze[b]	Royal Meillandina[b]	SCHobitet	5690	1984	Medium Yellow	The Conard-Pyle Co.
Scarlet Sunblaze[b]	Scarlet Meillandina[b]	MEIcubasi	4681	1982	Dark red	The Conard-Pyle Co.
Snow Sunblaze[b]		MEIghivon	8063	1991-NR	White	The Conard-Pyle Co.
Sunny Sunblaze[b]	Sunny Meillandina[b,e] Sunblaze[b]	MEIponal	6810	1985	Golden buff	The Conard-Pyle Co.
Sweet Sunblaze[b]	Pink Symphony[e] Pink Symphonie[b] Pretty Polly	MEItonje	PPAF	1987	Clear pink	The Conard-Pyle Co.
Sunset				1989-NR	Orange	

[a] NR = Not registered in *Modern Roses* 10 (Cairns, 1993) or not listed or listed as not registered in *Combined Rose List* 1994 (Dobson and Schneider, 1994).

[b] Some type of claim (i.e., Trademark) is made on the use of this name in commerce. This claim does not necessarily end with the expiration of the plant patent. The company listed as controlling propagation rights should be consulted about name usage.

[c] Trademark owner. Patent has expired.

[d] Plant Patent Applied For (PPAF).

[e] The registered name.

Table 2. American Association of Nurserymen rose grading standards.

GENERAL GARDEN ROSE

The standards specified apply only to field-grown garden roses when sold bare-root, or individually wrapped and packaged, or in cartons.

All grades of roses must have a well-developed root system and have proportionate weight and caliper according to grade and variety. Roses shall be graded by number and caliper of canes.

Rose bushes that do not meet these standards for the individual grades are defined as culls.

The grade sizes for each classification are minimum sizes, and not more than 10% of the rose plants in any bundle shall be below the size specified.

As used in the grade sizes below, "strong cane" means a cane that is healthy, vigorous, and fully developed so that it is hardened-off throughout. The caliper of the cane is measured not higher than 4 inches (10 cm) from the bud union.

HYBRID TEA, TEA, GRANDIFLORA, FLORIBUNDA, RUGOSA HYBRIDS, HYBRID PERPETUALS, MOSS, AND CLIMBING ROSES

Grade No. 1

At least 3 strong canes, $\frac{5}{16}$ inch (0.8 cm) in caliper and up, branched not higher than 3 inches (8 cm) from the bud union.

Grade No. 1½

At least 2 strong canes, $\frac{5}{16}$ inch (0.8 cm) in caliper and up, branched not higher than 3 inches (8 cm) from the bud union.

Grade No. 2

At least 2 canes, 1 of which shall be a strong cane, $\frac{5}{16}$ inch (0.8 cm) in caliper and up. The second shall be ¼ inch (0.6 cm) in caliper, branched not higher than 3 inches (8 cm) from the bud union.

Note: Although Floribunda roses are included in the above grade standard, it should be noted that Floribunda roses in this group will normally be, on the average, light for this class. Polyantha, Shrub, Landscape, and low growing Floribunda roses may be graded as below.

POLYANTHA, SHRUB, LANDSCAPE, AND LOW-GROWING FLORIBUNDA ROSES

Grade No. 1

At least 3 (strong) canes ¼ inch (0.6 cm) in caliper and up, branched not higher than 3 inches (8 cm) from the bud union.

Grade No. 1½

At least 2 (strong) canes, ¼ inch (0.6 cm) in caliper and up, branched not higher than 3 inches (8 cm) from the bud union.

Grade No. 2

At least 2 canes, one of which shall be a (strong) cane, ¼ inch (0.6 cm) in caliper and up.

FIELD GROWN MINIATURES (minimum standards)

Large Grower

Grade No. 1

At least 2 canes, of which 1 shall be ¼ inch (0.6 cm) in diameter and the other 9/32 inch (0.7 cm) in diameter, or 5 canes, one of which is ¼ inch (0.6 cm) in diameter, with 4 smaller healthy canes.

Grade No. 2

At least 2 canes, one of which is ¼ inch (0.6 cm) in diameter, plus 1 healthy cane.

Small Grower

Grade No. 1

At least 2 canes, 9/32 inch (0.7 cm) in diameter, or 5 small healthy canes.

Grade No. 2

2 healthy canes.

Root System

Grade No. 1

5 inches (13 cm) or more in length, spaced 50% or more around the shank in a balanced fashion.

Grade No. 2

3 to 5 inches (8–13 cm) in length, spaced 50% or more around the shank in a balanced fashion.

Table 3. A direct-stick, short-cycle production system used at a major Danish pot rose producer since 1988.

Direct-stick 4 cuttings per pot (4-inch; 10-cm).
Drench with fungicide directly following sticking.

Propagation in a fog system at 72°F (22°C) constant temperature with CO_2 enriched to 1200 ppm and HID lamps at 800 foot candles (105 µmol per second per square meter) for 16 hours per day.

Root initiation is in 14 days year-round.

- 4 days at 95% RH, 96% if sunny
- 4 days at 93% RH
- 3 days at 87% RH
- 3 days at 83% RH

After this rooting phase, the pots are automatically transported to the finishing greenhouses under the following conditions:

- Temperature and supplemental light the same as for the rooting phase.
- Fertilization using a solution with an electrical conductivity of 1.7 to 2.7 millimhos per cm from a complete balanced fertilizer plus iron.
- CO_2 enriched to 1000 ppm.

In the finishing greenhouses, root establishment requires 3 weeks during summer, 4 weeks in winter. They are then pinched, grown for 2 weeks, spaced to final spacing, and grown for 5 to 6 weeks for a total crop time of 12 to 14 weeks depending on the season.

Table 4. A typical program for outdoor propagation.

Facilities:

Usually a single poly-covered unheated quonset or a saran-shade house.

Time of Year:

June to August in the northeastern U.S. to year-round in southern Florida.

Special Requirements:

- Some shade, usually 30%, to cut the water loss due to the intense summer solar radiation. Never use more than 35% shade, even under high-light conditions.
- Heavier mist than indoors. The leaves should be kept wet at all times. Moisture stress outdoors is even less forgiving than in a greenhouse. Monitor the cuttings on the edges carefully, as they are most susceptible to heat/moisture stress.
- Use bigger, slightly hard cuttings. They are able to withstand stress better than young, very soft cuttings.
- Use a lighter soil mix to avoid water saturation in the pot due to heavier misting. However, make sure the soil mix is heavy enough to retain adequate moisture during subsequent winter forcing of the well-established plants that result from summer propagation outdoors.
- Bottom heat at 72°F (22°C), especially at night, if needed.
- Use a more concentrated rooting hormone, usually 0.2% IBA instead of 0.1%, and mix with a fungicide if one is not already present in the formulation used.
- Inspect closely for diseases such as black spot and downy mildew which are more prone to develop in these conditions. Spray labeled fungicides preventively for black spot and as needed for downy mildew.
- The hardening-off phase should have the following steps: reduce mist when roots begin to develop, stop the mist totally when roots are fully grown, and, finally, gradually remove the shade and/or the plastic.

Table 5. Two methods for forcing bare-root plants for Valentine's Day. (Adapted from Heins 1981).

METHOD I			METHOD II	
Temperature	**Procedure**	**Date**	**Procedure**	**Temperature**
35° to 40° (1.5° to 4.5°C)	Plants arrive	Nov. 1	—	—
48° to 50° (9° to 10°C)	Plant rose in pot and put in storage	Nov. 1 to 7	—	—
		Nov. 25 to 30	Plants arrive	35° to 40° (1.5° to 4.5°C)
60° to 62°F (15.5° to 16.5°C)	Start forcing[a]	Dec. 1 to 7	Plant and start forcing[a]	60° to 62°F (15.5° to 16.5°C)
	Remove dominant shoots	Dec. 15 to 25	Remove dominant shoots	
60° to 65°F (16° to 18°C)	Buds color, market	Feb. 5 to 10	Buds color, market	60° to 65°F (16° to 18°C)

[a] 250 foot candles supplemental light for a minimum of 12 hours per day in low-winter-light climates.

Table 6. Conversions of photometric to radiant energy measurements. Adapted from Thimijan and Heins (1983) and Tsujita (1987).

PHOTOMETRIC		RADIANT ENERGY (400–700 NM)			
		µMOL PER SECOND PER SQUARE METER		WATTS PER SQUARE METER	
FOOT CANDLES	LUX	DAYLIGHT	HPS	DAYLIGHT	HPS
300	3,240	60	39.5	13.1	7.9
400	4,320	80	52.7	17.5	10.6
500	5,400	100	65.9	21.9	13.2
600	6,480	120	79.0	26.3	15.8
700	7,560	140	92.2	30.6	18.5
800	8,640	160	105.4	35.0	21.2
900	9,720	180	118.5	39.4	23.8
1000	10,800	200	132.7	43.8	26.6

Conversion Factors

1. Footcandle to lux—multiply footcandle readings by 10.8.

2. Lux to µmol per second per square meter PAR (400–700 nm)—divide lux reading by the following factor for each lamp or light source:

sun and sky, daylight	54
high pressure sodium (HPS)	82
metal halide	71
cool white fluorescent	74

3. µmol per second per square meter (µE per second per square meter) to Watts per square meter photosynthetically active radiation (PAR) (400–700 nm)—divide µmol per second per square meter reading by the following factor for each lamp or light source:

sun and sky, daylight	4.57
high pressure sodium (HPS)	4.98
metal halide	4.59
cool white fluorescent	4.59

Table 7. Relation of days to flower and bud diameter for flowers on Garnette plants forced at 61°F night/68°F day (16°C night/20°C day) temperature. Values were computed using the formula days to flower = 31.6–[(2.1) (bud diameter in mm)]. Adapted from Heins (1981).

BUD DIAMETER		DAYS TO
MM	INCHES	FLOWER
1	0.04	30
2	0.08	27
3.2	0.125	25
4	0.16	23
6.35	0.25	18
8	0.31	15
10	0.39	11
12.7	0.50	5

Table 8. Key to rose nutrient disorders. Adapted from White (1987).

A. Older plant parts affected first.

B. Older leaves turn light green to completely yellow or entire leaf is chlorotic but usually remains on the plant. Growth is reduced.

C. Light green to completely yellow leaves remain on plant. There is a failure of buds to develop properly. Stems weak and spindly. Small flowers of light color are common.

NITROGEN (N) DEFICIENCY

CC. Entire leaf is chlorotic (yellow between veins). Plants are stunted with large, necrotic white areas symmetrically distributed on both sides of midrib of the leaflets between larger lateral veins of the older leaves; leaf edges of older leaves cup down.

MAGNESIUM (Mg) DEFICIENCY

BB. Older leaves are not as above.

C. Older foliage drops without turning yellow.

D. The leaves are dull grey-green in color and the buds slow to develop; leaf edges of older leaves may cup down.

PHOSPHORUS (P) DEFICIENCY

DD. Older leaves drop rapidly—-plants may need to be pinched to keep foliage; necrotic leaf margins.

SULFUR (S) TOXICITY, ROOT LOSS, SPRAY INJURY, AMMONIUM TOXICITY, AIR POLLUTION

CC. Older foliage does not drop at first and visual symptoms are apparent in localized areas on the leaves.

D. Water-soaked areas appear along the mid-vein or along other main veins, then these areas remain green as the rest of the leaf turns first yellow, then brown; leaves eventually drop after many cup down.

ZINC (Zn) TOXICITY

DD. Margins of leaf are affected first.

E. Margins of leaf are yellow or brown.

F. Margins become yellow then turn brown, leaves sometimes become purple. Young shoots become hardened and stunted, flower buds may abort and abscise.

POTASSIUM (K) DEFICIENCY

FF. Browning of leaf serrations of leaflets, the dead tissue is separated from the green tissue by a distinctive pink margin; brown, irregular spots may develop followed by necrosis and eventual leaf drop.

BORON (B) TOXICITY

EE. Purplish-brown blotches appear along margin, the foliage wilts, and little new growth occurs.

CHLORIDE (Cl) TOXICITY

AA. Younger plant parts affected first.

B. Chlorosis of young leaves.

C. Terminal chlorosis most obvious symptoms.

D. Failure of buds to develop after cut or pinch.

NITROGEN (N) EXCESS

DD. Interveinal areas yellow, also general stunting of root system.

IRON (Fe) DEFICIENCY

CC. Net-veined leaves or older leaves with black spots most distinguishing symptoms.

D. Interveinal areas yellow but smallest veins green, more of a netted appearance.

Table 8 continued.

MANGANESE (Mn) DEFICIENCY

DD. Small black spots on older leaves. Also may cause iron deficiency symptoms to appear.

MANGANESE (Mn) EXCESS

BB. Shoots not chlorotic but often die or are hard.

C. New growth ceases or dies.

D. Buds dead, leaves do not develop or are distorted and internodes shortened. Witch's broom appearance.

BORON (B) DEFICIENCY

DD. Young leaves develop light edges; apical meristem dies, resulting in development of many small side branches. Possible to confuse with magnesium deficiency.

COPPER (Cu) DEFICIENCY

DDD. New growth dies without development of witch's broom appearance and plants become defoliated. Many brown and dead roots.

E. On older leaves, leaf edges cup down.

CALCIUM (Ca) DEFICIENCY

EE. On older leaves, leaf edges do not cup down. New shoot growth aborted, distorted, strap leaves.

ZINC (Zn) DEFICIENCY

Table 9. Soil test interpretation key for Modified Morgan and Spurway analyses. Adapted from White, 1987. The Modified Morgan analysis is used at Pennsylvania State University. See text for explanation.

	MODIFIED MORGAN	SPURWAY
Nitrogen (NO_3)	100–400 ppm[a]	25–100 ppm[b]
Phosphorus (P)	400–900 ppm	4–6 ppm
Potassium (K)	2.0–4.0 me/100 g and 4.5% saturation	30–50 ppm
Calcium (Ca)	1.5–20 me/100 g and 60–70% saturation	over 200 ppm
Magnesium (Mg)	4.10 me/100 g and 9–10% saturation	
pH	5.5–6.5	4.5–7.0
Total soluble salts	100–300 (2:1 water-soil)	150 (2:1)

[a] ppm = parts per million, parts of dry soil.
[b] ppm = parts per million, parts of extract solution.

Table 10. Leaf analysis interpretation key for roses. Adapted from White, 1987.

ELEMENT	OERTLI (1966)[a] DEFICIENT	HEALTHY	BOODLEY & WHITE (1969)[b] STANDARD RANGE	CARLSON (1966) OPTIMAL	SADASIVAIAH &[c] HOLLEY (1973) NORMAL RANGE
			Percent (%)		
Nitrogen	1–1.5	4–6	3.0–5.0	3.2–4.0	3.00–3.50
Phosphorus	0.01–0.03	0.2 or more	0.2–0.3	0.2–0.3	0.28–0.32
Potassium	0.3–1.0	1.0 or more	1.8–3.0	1.5–1.8	2.00–2.50
Calcium		1.0–3.5	1.0–1.5	1.0–1.9	1.00–1.60
chlorotic zone	0.1				
green zone	0.4				
Magnesium	0.1–0.2	0.2 or more	0.25–0.35	0.28–0.34	0.28–0.32
Sulfur					0.16–0.21
			Parts per million (ppm)		
Zinc			15–50	40	20–40
Manganese			30–250	300–900	70–120
Iron		60 or more	50–150	80–100	80–120
moderate chlorosis	40–60				
severe chlorosis	20–30				
Copper	10–20	20 or more	5–15	10–14	7–15
Boron	0.03–0.05	0.1 or more	30–60	20–40	40–60

[a] Based on studies in solution culture using Red Delight and Better Times cut-roses; plant samples were tissue showing deficiency symptoms or healthy tissue. Deficient plants were showing definite symptoms of deficiency. Healthy plants appeared to be growing normally and without symptoms of deficiencies or toxicities.

[b] Based on observations and surveys of commercial greenhouse cut-roses grown in soil mixtures; plant samples were the uppermost 5-leaflet leaf on a stem with a flower bud just beginning to show color. The amount is sufficiently above the critical level so that any change in other elements should not induce a deficiency through antagonistic action or by other means. When below the critical level or minimum requirement for the element, a deficiency probably exists. Reduction in growth, yield, or quality may be readily apparent.

[c] Based on Forever Yours cut-roses produced in granitic gravel and irrigated with nutrient solutions 1 or more times daily. The 2 uppermost 5-leaflet leaves on a flowering stem nearing the harvest stage were used for spectrographic or chemical analysis.

Table 11. Flower development stages of miniature roses (published with permission from Cushman et al., 1994).

FLOWER STAGE	DESCRIPTION
1	Tight bud, calyx not reflexing
2	Showing color, calyx reflexing, no petals reflexed
3	Full color, petals beginning to reflex, traditional bud stage
4	Several petals reflexed, traditional exhibition stage
5	Full open
6	Postharvest life over, petal wilting or abscission

Table 12. Industry standards for potted plant pack sizes from the Society of American Florists and the Produce Marketing Association.

POT DIAMETER	NUMBER OF POTS
76 mm (3.0 in)	28
102 mm (4.0 in)	15
114 mm (4.5 in)	15
127 mm (5.0 in)	10
140 mm (5.5 in)	8
152 mm (6.0 in)	6
165 mm (6.5 in)	6
178 mm (7.0 in)	4
191 mm (7.5 in)	4
203 mm (8.0 in)	4
216 mm (8.5 in)	4
229 mm (9.0 in)	3
254 mm (10 in)	2
357 mm (14 in)	1

Materials:

- Minimum 1724 kPa (250 lb/in^2) bursting test fiberboard
- 161 g/m^2 (88 lb/1000 ft^2) corrugated medium
- C Flute corrugation 42 flutes per 0.3 m (linear ft), 3.5 mm ($\frac{9}{64}$ in) high
- Waterproof adhesive
- Fiberboard dividers for long distance transport or high relative humidity
- Moisture-resistant wax or plastic-impregnated fiberboard tray
- Cut-out hand grips for ease in handling

Table 13. Pot rose production budget.

GENERAL PRODUCTION AND ASSET INPUTS		
	NUMBER OF UNITS	COST/VALUE PER UNIT
Analysis year	1990	—
Long-term interest rate	11%	—
Taxes, ins. rate	0.1	—
House sq. ft.	20000	—
Bench sq. ft.	15400	—
Land (includes prep.)		
(acres and $/acre)	3	$2000
Buildings—greenhouse	1	$60000
Plastic	1	$3000
Service buildings	1	$20000
Paving	1	$2000
Cooler	1	$3000
Benches	1	$20000
Greenhouse floor	1	$4000
Irrigation system	1	$12000
Wiring	1	$8500
Equipment		
Fertilizer	1	$1400
Spray	1	$800
Office	1	$1400
Lunch Room	1	$800
Shop	1	$2000
Intern transp.	1	$1100
Truck	1	$16000
Emergency equipment	1	$2500
Shade cloth	1	$2000
Miscellaneous	1	$4000

Table 14. Pot rose production budget.

ASSET YEARS OF LIFE	
Buildings—greenhouse	20
Plastic	2
Service buildings	25
Paving	15
Cooler	20
Benches, built	5
Greenhouse floor	5
Irrigation system	13
Wiring	20
Equipment	
Fertilizer	7
Spray	7
Office	10
Lunch Room	10
Shop	7
Internal transportation	10
Truck	7
Emergency equipment	15
Shade cloth	5
Miscellaneous	5

Table 15. Pot rose production budget.

OUTPUT INFORMATION: INVESTMENT AND FIXED COST ESTIMATES								
ITEM	NUMBER	PER UNIT VALUE	NEW INVESTMENT	AVERAGE INVESTMENT	YEARS OF LIFE	ANNUAL DEPRECIATION	INTEREST INVESTMENT	TOTAL
Buildings—greenhouse	1	$60,000	$60,000	$30,000	20	$1500	$3,300	$4800
Plastic	1	$ 3,000	$3,000	$1,500	2	$750	$165	$915
Service buildings	1	$20,000	$20,000	$10,000	25	$400	$1,000	$1400
Paving	1	$2,000	$2,000	$1,000	15	$67	$110	$177
Cooler	1	$3,000	$3,000	$1,500	20	$75	$165	$240
Benches, built	1	$20,000	$20,000	$10,000	5	$2000	$1,100	$3100
Greenhouse floor	1	$4,000	$4,000	$2,000	5	$400	$220	$620
Irrigation system	1	$12,000	$12,000	$6,000	13	$462	$660	$1122
Wiring	1	$8,500	$8,500	$4250	20	$213	$468	$680
Equipment								
Fertilizer	1	$1,400	$1,400	$700	7	$100	$77	$177
Spray	1	$800	$800	$400	7	$57	$44	$101
Office	1	$1,400	$1,400	$700	10	$70	$77	$147
Lunch Room	1	$800	$800	$400	10	$40	$44	$84
Shop	1	$2,000	$2,000	$1,000	7	$143	$110	$253
Internal transportation	1	$1,100	$1,100	$550	10	$55	$61	$116
Truck	1	$16,000	$16,000	$8,000	7	$1143	$880	$2023

ITEM	NUMBER	PER UNIT VALUE	NEW INVESTMENT	AVERAGE INVESTMENT	YEARS OF LIFE	ANNUAL DEPRECIATION	INTEREST INVESTMENT	TOTAL
Emergency equipment	1	$2,500	$2,500	$1,250	15	$83	$138	$221
Shade cloth	1	$2,000	$2,000	$1,000	5	$200	$110	$310
Miscellaneous	1	$4,000	$4,000	$2,000	5	$400	$220	$620
Total			$164,500	$82,250		$8,157	$8,948	$17,104
Land (total value of land and clearing costs; long-term interest rate)							$660	
Total (including land) annual building/equipment cost								$17,764
Buildings/equipment new investment				$164,500				
Buildings/equipment average investment				$82,250				
Buildings/equipment total annual cost				$17,764				
Total annual building/equipment cost per sq. ft greenhouse				$0.89				
Total annual building/equipment cost per sq. ft. bench space				$1.15				

Table 16. Pot rose production budget.

ANNUAL OVERHEAD COST ESTIMATE

	TOTAL ANNUAL COST	% OF TOTAL OVERHEAD	% OF OVERHEAD & FIXED COST
Salaries	$30,000	46.50%	36.80%
Payroll tax	$2,200	3.40%	2.70%
Unemployment compensation	$340	0.50%	0.40%
Workman's compensation	$450	0.70%	0.60%
Insurance	$3,000	4.70%	3.70%
Water & sewer	$600	0.90%	0.70%
Telephone	$1,300	2.00%	1.60%
Fuel oil	$11,000	17.10%	13.50%
Electricity	$3,000	4.70%	3.70%
Repairs & maintenance	$3,000	4.70%	3.70%
Property tax	$500	0.80%	0.60%
Advertising	$500	0.80%	0.60%
Truck expense	$6,000	9.30%	7.40%
Bad debt	$1,000	1.60%	1.20%
Professional fees	$500	0.80%	0.60%
Supplies	$300	0.50%	0.40%
Miscellaneous	$800	1.20%	1.00%
Total annual overhead costs			$64,490
Per sq. ft. greenhouse			$3.22
Per sq. ft. bench space			$4.19
Total annual overhead and building/equipment cost			
Per sq. ft. greenhouse			$4.11
Per sq. ft. bench space			$5.34

Table 17. Pot rose production budget.

POTTED ROSE PLANT VARIABLE COST COMPUTATION

A. *General assumptions variables*

Number of plants	2000
Percentage loss	5%
Planting week number	12
First spacing (inch)	4.5
Second spacing	8
Number of weeks at 1st spacing	2
Number of weeks at 2nd spacing	6
Planned mark-up (%)	30%

B. *Calculated values*

Plants available for sale	1900
Sq. ft. bench, 1st spacing	267.2
Sq. ft. bench, 2nd spacing	844.4
Bench space, 1st spacing (%)	2%
Bench space, 2nd spacing (%)	5%
Fixed cost, 1st spacing	$0.03
Fixed cost, 2nd spacing	$0.22
Total fixed cost per pot	$0.24

C. *Variable cost estimate*

NAME	UNIT	COST PER UNIT	QUANTITY	TOTAL COST	COST PER CONTAINER
MATERIAL COST:					
Containers (pots)	Each	$0.12	2000	$240.00	$0.12
Media (soil)	Cu ft	$2.50	40	$100.00	$0.05
Cuttings/seed	Each	$0.78	2000	$1,560.00	$0.78
Packing					
(sleeves, boxes)	Thou	$60.00	2	$120.00	$0.06
Fertilizer:					
Water solubles	Lbs	$1.00	50	$50.00	$0.03
Chemicals:					
Fungicide	Oz	$1.19	6	$7.14	$0.00
Fungicide	Oz	$1.17	10	$11.70	$0.01
Insecticide	Oz	$1.56	4	$6.24	$0.00
Insecticide	Oz	$0.56	4	$2.24	$0.00

Table 17 continued.

NAME	UNIT	COST PER UNIT	QUANTITY	TOTAL COST	COST PER CONTAINER
Special labor:					
Pinching-pruning	Hr	$5.50	10	$55.00	$0.03
Spraying pesticide	Hr	$5.50	7	$38.50	$0.02
Packaging/sleeving	Hr	$5.50	12	$66.00	$0.03
Transporting	Hr	$5.50	4	$22.00	$0.01
Repotting	Hr	$5.50	20	$110.00	$0.05
Miscellaneous	Hr	$5.50	25	$110.00	$0.06
Total				$2,497.70	$1.25

D. *Cost summary and sale price projection*

	TOTAL COST	COST PER POT
Variable cost	$2,498	$1.25
Fixed & overhead cost	$463	$0.24
Fixed & variable cost	$2,962	$1.49
Estimated sale pice		$1.94

Table 18. Pot rose production budget.

ADDITIONAL VARIABLE COST FOR SUPPLEMENTAL LIGHTING WITH HIGH-INTENSITY DISCHARGE (HID) FIXTURES

A. *General assumptions variables*

Fixture type	400 watt sodium vapor HID
Light intensity	500 foot candles
Duration	16 hours per day
Mounting height	5 ft above crop canopy
Fixture spacing[a]	5.84 watts per sq. ft.
1st spacing (2 weeks)—pots per sq. ft.	7.11
2nd spacing (6 weeks)—pots per sq. ft.	2.25
Electric usage per sq. ft. greenhouse space	0.0934 kilowatt hours per day

B. *Calculated variable cost estimate*

	Electric cost ($ per kilowatt hour)			
	0.026	0.050	0.075	0.100
Electric cost per sq. ft. Greenhouse space ($ per day)	0.0024	0.0047	0.0070	0.0090
Electric cost per sq. ft. Bench space ($ per day)	0.0031	0.0061	0.0090	0.0116
Electric cost for 1st spacing ($ per pot)	0.0061	0.0120	0.0177	0.0228
Electric cost for 2nd spacing ($ per pot)	0.0579	0.1139	0.1680	0.2165
Total electric cost ($ per pot)	0.0640	0.1259	0.1857	0.2393

[a] Spacing will vary according to greenhouse design and bench layout. Consult fixture manufacturer.

References

Andersson, N. E. 1991. The influence of constant and diurnally changing CO_2 concentrations on plant growth and development. J. Hort. Sci. 66:569–574.

Anonymous. 1987. Greenhouse operating survey report. Bedding Plants, Inc. 69 pp.

Anonymous. 1990. American standards for nursery stock. American Association of Nurserymen Inc., Washington, D.C. 21–22.

Armitage, A. M., and M. J. Tsujita. 1979. Supplemental lighting and nitrogen nutrition effects on yield and quality of Forever Yours roses. Can. J. Plant Sci. 59:343–350.

Asaoka, M., and R. D. Heins. 1982. Influence of supplemental light and preforcing storage treatments on the forcing of 'Red Garnette' rose as a pot plant. J. Amer. Soc. Hort. Sci. 107:548–552.

Boyd, R. M., T. D. Phillips, T. M. Blessington, and S. P. Myers. 1982. Costs of producing selected floricultural crops. AEC Research Report No. 133. Mississippi Agricultural and Forestry Experiment Station. 62 pp.

Boyd, R. M., and T. D. Phillips. 1983. Programming greenhouse floricultural crops. AEC Research Report No. 144. Mississippi Agricultural and Forestry Experiment Station. 52 pp.

Brickell, C. D. 1980. International Code of Nomenclature for Cultivated Plants—1980. The International Bureau for Plant Taxonomy and Nomenclature. Utrecht, Netherlands.

Brumfield, R. G., P. V. Nelson, A. J. Coutu, D. H. Willits, and R. S. Sowell. 1981. Overhead costs of greenhouse firms differentiated by size of firm and market channel. Tech. Bull. No. 269. N. C. Agricultural Research Service. 89 pp.

Cairns, T., ed. 1993. Modern Roses 10. The American Rose Society. Shreveport, Louisiana. 740 pp.

Cameron, A. C., M. S. Reid, and G. W. Hickman. 1981. Using STS to prevent flower shattering in potted flowering plants-progress report. Flower and Nursery Report—University Calif., Fall:1–3.

Carlson, W. H. 1966. Foliar analysis, a new tool. Roses Inc. Bull., December:25–28.

Carpenter, W. J., and G. A. Anderson. 1972. High intensity supplementary lighting increases yields of greenhouse roses. J. Amer. Soc. Hort. Sci. 97:331–334.

Chen, L. C. 1990. Growth regulator reversal of simulated high temperature shipping effects on flower senescence and leaf abscission in miniature potted rose plants. M.S. Thesis. Texas A&M University. College Station, TX.

Clark, D. G. 1990. Postharvest handling of potted roses. M.S. Thesis. Clemson University. Clemson, SC.

Clark, D. G., J. W. Kelly, and D. R. Decoteau. 1991a. Influence of end-of-day red and far-red light on potted roses. J. Environ. Hort. 9(3):123–127.

Clark, D. G., J. W. Kelly and H. B. Pemberton. 1991b. Postharvest quality characteristics of cultivars of potted rose in response to holding conditions and cytokinins. HortScience 26(9):1195–1197.

Clark, D. G., J. W. Kelly, and N. C. Rajapakse. 1993. The effects of carbon dioxide enrichment on production and postharvest characteristics of *Rosa hybrida* L. \`Meijikatar'. J. Amer. Soc. Hort. Sci. 118:613–617.

Cushman, L. C., H. B. Pemberton, and J. W. Kelly. 1994. Cultivar, flower stage, silver thiosulfate, and BA interactions affect performance of potted miniature roses. HortScience 29:805–808.

Dobson, B. R., and P. Schneider. 1994. Combined Rose List. Peter Schneider. Mantua, Ohio. 157 pp.

Elliott, W. H. 1991. Property rights and plant germplasm. HortScience 26:364–365.

Faber, W. R., W. M. Brooks, H. K. Tayama, J. C. Peterson, J. L. Robertson, C. C. Powell, and R. K. Lindquist. 1986. Tips on growing bedding plants. Ohio Cooperative Extension Service. The Ohio State University. 24 pp.

Fjeld, T., H. R. Gislerød, V. Revhaug, and L. M. Mortensen. 1994. Keeping quality of cut roses as affected by high supplementary irradiation. Scientia Hort. 57:157–164.

Gudin, S. 1992. Effect of preharvest growing temperatures on the development of cut roses. Postharvest Biol. and Technol. 2:155–161.

Halevy, A. H., and A. M. Kofranek. 1976. The prevention of flower bud and leaf abscission in pot roses during simulated transport. J. Amer. Soc. Hort. Sci. 101:658–660.

Hammer, P. A. 1992. Other flowering pot plants. In Introduction to Floriculture, ed. R. A. Larson. Academic Press, Inc. San Diego. 477–509.

Hardenburg, R. E., A. E. Watada, and C. Y. Wang. 1986. The commercial storage of fruits, vegetables, and florist and nursery stocks. USDA Agric. Handb. 66:100–101.

Hartmann, H. T., D. E. Kester, and F. T. Davies, Jr. 1990. Plant propagation principles and practices. 5th edition. Prentice Hall. Englewood Cliffs, New Jersey.

Heins, R. D. 1981. Forcing pot roses for Valentine's Day. Florists' Review 169 (4376/15 Oct.):14–15.

Hicklenton, P. R. 1988. CO_2 Enrichment in the Greenhouse: Principles and Practice. Growers Handbook Series Vol. 2. A. M. Armitage, General Ed. Timber Press. Portland Oregon. 58 pp.

Higginbotham, J. S. 1992. Cracking the code. American Nurseryman 176(11):36–49.

Horridge, J. S., and K. E. Cockshull. 1974. Flower initiation and development in the glasshouse rose. Scientia Hort. 2:273–284.

Horst, R. K. 1983. Compendium of rose diseases. American Phytopathological Society. St. Paul, Minn. 50 pp.

Hughes, H. E., and J. J. Hanan. 1978. Effect of salinity in water supplies on greenhouse rose production. J. Amer. Soc. Hort. Sci. 103:694–699.

Jiao, J., X. Wang, and M. J. Tsujita. 1990. Whole plant net photosynthesis of miniature roses influenced by light, CO_2, and temperature. Acta Hort. 272:261–265.

Kelly, J. W. 1990. A crop budget for potted greenhouse roses. S. C. Greenhouse Growers' Association Newsletter.

Kelly, J., J. Rathwell, R. Sutton, and D. Luke. 1990. Greenhouse enterprise budget calculator. Clemson University Cooperative Extension Service. 8-pp. brochure and software.

Khayat, E., and N. Zieslin. 1982. Environmental factors involved in the regulation of sprouting of basal buds in rose plants. J. Exp. Bot. 33:1286–1292.

Khosh-Khui, M., and R.A.T. George. 1977. Responses of glasshouse roses to light conditions. Scientia Hort. 6:223–235.

Kyalo, T. M. 1992. Optimization of production quality and postproduction longevity for miniature pot roses. M.S. Thesis. Texas A&M University. College Station, TX.

Kyalo, T. M., H. B. Pemberton, and J. M. Zajicek.1996. Seasonal growing environment affects quality characteristics and postproduction longevity of potted miniature roses. HortScience 31:120–122.

Kyalo, T. M., H. B. Pemberton, J. M. Zajicek, and G. V. McDonald. 1993. Effect of uniconazole rate and time of application on morphological quality characteristics of miniature pot roses. HortScience 28:548 (Abstr.)

Laurie, A., D. C. Kiplinger, and K. S. Nelson. 1969. Commercial Flower Forcing. 7th ed. McGraw-Hill Book Co. New York, NY.

MacKay, I. 1985. Roses—the spring pot crop. In Ball Red Book, 14th ed., ed. V. Ball. Reston Publishing Company, Inc. Reston, VA. 677–685

Mastalerz, J. W. 1987. Environmental factors light, temperature, carbon dioxide. In Roses, A Manual on the Culture, Management, Diseases and Insects of Greenhouse Roses, ed. R. W. Langhans. Roses Incorporated. Haslett, MI.

———. 1977. The Greenhouse Environment. John Wiley & Sons. New York. 629 pp.

Maxie, E. C., R. F. Hasek, and R. H. Sciaroni. 1974. Keep potted roses cool. Calif. Flower and Nursery Rep., March:9–10.

McCann, K. R. 1991. Mini potted roses hit the big time. Greenhouse Grower 9 (No. 13):20–23.

McMahon, M. J., and J. W. Kelly. 1990. Influence of spectral filters on height, leaf chlorophyll, and flowering of *Rosa* x *hybrida* 'Meirutral'. J. Environ. Hort. 8:209–211.

Miller, R. 1987. Tips for the mini pot rose grower. GrowerTalks 51(5-September):80–82, 85.

Moe, R. 1970. Growth and flowering of potted roses as affected by temperature and growth retardants. Meld. Nor. Landbrukshoegsk 49:1–16.

———. 1972. Effect of daylength, light intensity, and temperature on growth and flowering in roses. J. Amer. Soc. Hort. Sci. 97: 796–800.

———. 1973. Propagation, growth and flowering of potted roses. Acta Hort. 31:35–50.

———. 1975. The effect of growing temperature on keeping quality of cut roses. Acta Hort. 41:77–92.

Moe, R., and T. Kristoffersen. 1969. The effect of temperature and light on growth and flowering of *Rosa* 'Baccara' in greenhouses. Acta Hort. 14:157–167.

Monteiro, J. A., T. A. Nell, and J. E. Barrett. 1992. Potted miniature rose longevity affected by flower respiration. HortScience 27:603 (Abstr.)

Mor, Y., and N. Zieslin. 1987. Plant growth regulators in rose plants. Hort. Rev. 9:53–73.

Mor, Y., A. H. Halevy, A. M. Kofranek, and J. Kubota. 1986. Forcing pot roses from own-root cuttings; effect of growth retardants and light. Acta Hort. 189:201–208.

Mortensen, L. M. 1991. Effects of temperature, light and CO_2 level on growth and flowering of miniature roses. Nor. J. Agri. Sci. 5:295–300.

Nell, T. A., and C. V. Noordegraaf. 1991. Simulated transport, postproduction irradiance influence postproduction performance of potted roses. HortScience 26:1401–1404.

———. 1992. Postproduction performance of potted rose under simulated transport and low irradiance levels. HortScience 27:239–241.

Oertli, J. J. 1966. Nutrient deficiencies in rose plants. Roses Inc. Bull., June 1968, and Florists' Review 138 (3578–3585).

Parsons, S. B. 1869. Parsons on the Rose. Orange Judd and Company. New York. p. 116.

Pemberton, H. B., and W. E. Roberson. 1990. Critical tissue levels needed for maintenance of rose plant viability. HortScience 25:863 (Abstr.).

Pobudkiewicz, A., and K. L. Goldsberry. 1989. Controlling the growth habit of dwarf pot roses with uniconazole (Sumagic™). Colorado Greenhouse Growers' Assoc. Res. Bull. 471:1–2.

Post, K. 1949. Florist Crop Production and Marketing. Orange Judd Publishing Company. New York.

Pryor, M., C. Harwood, and H. Wessig. 1987. Plant production. In Roses, A Manual on the Culture, Management, Diseases and Insects of Greenhouse Roses, ed. R. W. Langhans. Roses Incorporated. Haslett, Michigan.

Rajapakse, N. C., and J. W. Kelly. 1994. Influence of spectral filters on growth and postharvest quality of potted miniature roses. Scientia Hort.56:245–256.

Reid, M. S., J. L. Paul, M. B. Farhoomand, A. M. Kofranek, and G. L. Staby. 1980. Pulse treatments with the silver thiosulfate complex extend the vase life of cut carnations. J. Amer. Soc. Hort. Sci. 105:25–27.

Sachs, R. M., A. M. Kofranek, and W. P. Hackett. 1976. Evaluating new pot plant species. Florists' Review 159(4116):35–36, 80–84.

Sadasivaiah, S. P., and W. D. Holley. 1973. Ion balance in nutrition of greenhouse roses. Roses Inc. Bull. Suppl. November:1–27.

Serek, M. 1993. Ethephon and silver thiosulfate affect postharvest characteristics of *Rosa* hybrida 'Victory Parade'. HortScience 28:199–200.

Serek, M., and A. S. Andersen. 1993. AOA and BA influence on floral development and longevity of potted 'Victory Parade' miniature rose. HortScience 28:1039–1040.

Shafer, B. S. 1985. Influence of cultivar, price, and longevity on consumer preferences for potted chrysanthemums. M.S. Thesis. Texas A&M University. College Station, TX.

Smith, R. B., and L. J. Skog. 1992. Cold and controlled atmosphere storage of miniature potted roses. Ontario Ministry of Agriculture and Food. Horticultural Research Institute of Ontario. Vineland Station, Ontario Canada. Report for 1991–92. 26 pp.

Staby, G. L., J. L. Robertson, D. C. Kiplinger, and C. Conover. 1976. Chain of Life. Ohio Flower Assoc. Hortic. Ser. No. 432.

Tjosvold, S. A., M. Wu, and M. S. Reid. 1994. Reduction of postproduction quality loss in potted miniature roses. HortScience 29:293–294.

Tsujita, M. J. 1987. High intensity supplementary radiation of roses. In Roses, A Manual on the Culture, Management, Diseases and Insects of Greenhouse Roses, ed. R. W. Langhans. Roses Incorporated. Haslett, Michigan.

Vaut, G. A., R. L. Christensen, T. C. Stone, and J. F. Smiarouski. 1973. Greenhouse linear programming. Cooperative Extension Service. Univ. Massachusetts. Publication No. 93.

White, J. W. 1987. Fertilization. In Roses, A Manual on the Culture, Management, Diseases and Insects of Greenhouse Roses, ed. R. W. Langhans. Roses Incorporated. Haslett, MI.

Warncke, D. D. and D. M. Krauskopf. 1983. Greenhouse growth media: testing and nutrition guidelines. Cooperative Extension Service. Michigan State University. Extension Bulletin E-1736.

White, J. W., and E. J. Holcomb. 1987. Water requirements and irrigation practices. In Roses, A Manual on the Culture, Management, Diseases and Insects of Greenhouse Roses, ed. R. W. Langhans. Roses Incorporated. Haslett, MI.

Zieslin, N., and A. H. Halevy. 1975. Flower bud atrophy in 'Baccara' roses. II. The effect of environmental factors. Scientia Hort. 3:383–391.

Zieslin, N. and R. Moe. 1985. Rosa. In Handbook of Flowering. Vol. 4. pp. 214–225, ed. A.H. Halevy. CRC Press. Boca Raton, FL.

Zieslin, N., and Y. Mor. 1990. Light on roses. A review. Scientia Hort. 43:1–14.

Zieslin, N., and M. J. Tsujita. 1990a. Response of miniature rose to supplementary illumination. 1. Light intensity. Scientia Hort. 42:113–121.

———. 1990b. Response of miniature rose to supplementary illumination. 2. Effect of stage of plant development and cold storage. Scientia Hort. 42:123–131.

Zieslin, N., A. H. Halevy, and I. Biran. 1973. Sources of variability in greenhouse rose flower production. J. Am. Soc. Hortic. Sci. 98:321–324.